Advances in Seafood Biochemistry: Composition and Quality Seafood Biochemistry

Advances in Seafood Biochemistry: Composition and Quality Seafood Biochemistry

Balaji Yadav Maddina

Advances in Seafood Biochemistry: Composition and Quality Seafood Biochemistry

ISBN 978-93-5111-795-7

Published in 2016 in India by

RANDOM PUBLICATIONS

4376-A/4B, Gali Murari Lal, Ansari Road
New Delhi-110 002
Phone : +9111-43580356, 011-23289044, 011-43142548
e-mail: sales@randompublications.com,
info@randompublications.com, randomexports@gmail.com

Reprinted 2024

Type Setting by : Friends Media, Delhi-110089
Digitally Printed at : Replika Press Pvt. Ltd.

Preface

Seafood is any form of sea life regarded as food by humans. Seafood prominently includes fish and shellfish. Shellfish include various species of molluscs, crustaceans, and echinoderms. Historically, sea mammals such as whales and dolphins have been consumed as food, though that happens to a lesser extent in modern times. Edible sea plants, such as some seaweeds and microalgae, are widely eaten as seafood around the world, especially in Asia. In North America, although not generally in the United Kingdom, the term "seafood" is extended to fresh water organisms eaten by humans, so all edible aquatic life may be referred to as seafood. For the sake of completeness, this article includes all edible aquatic life. The harvesting of wild seafood is usually known as fishing or hunting, and the cultivation and farming of seafood is known as aquaculture, or fish farming in the case of fish. Seafood is often distinguished from meat, although it is still animal and is excluded in a strict vegetarian diet. Seafood is an important source of protein in many diets around the world, especially in coastal areas. Most of the seafood harvest is consumed by humans, but a significant proportion is used as fish food to farm other fish or rear farm animal. Some seafoods are used as food for other plants. In these ways, seafoods are indirectly used to produce further food for human consumption. Products, such as fish oil and spirulina tablets are also extracted from seafoods. Some seafood is feed to aquarium fish, or used to feed domestic pets, such as cats, and a small proportion is used in medicine, or is used industrially for non-food purposes.

– Author

Contents

1

Basic Principles of Food Process Engineering

As an introduction to food process engineering, this book describes the scientific principles on which food processing is based and gives some examples of the application of these principles in several food industries. After understanding some of the basic theory, students should study more detailed information about the individual industries and apply the basic principles to their processes. For example, after studying heat transfer in this book, the student could seek information on heat transfer in the canning and freezing industries. To supplement the relatively few books on food process engineering, other sources of information are used, for example:

- *Specialist descriptions of particular food industries:* These in general are written from a descriptive point of view and deal only briefly with engineering.
- *Textbooks in chemical and biological process engineering:* These are studies of processing operations but they seldom have any direct reference to food processing. However, the basic unit operations apply equally to all process industries, including the food industry.
- *Engineering handbooks*: These contain considerable data including some information on the properties of food materials.
- *Periodicals:* In these can often be found the most up-to-date information on specialized equipment and processes, and increased basic knowledge of the unit operations.

The study of process engineering is an attempt to combine all forms of physical processing into a small number of basic operations, which are called unit operations.seafood processes may seem bewildering in their diversity, but careful analysis will show that these complicated and differing processes can be broken down into a small number of unit operations. For example, consider heating of which innumerable instances occur in every food industry. There are many reasons for heating and cooling–for example, the baking of bread, the freezing of meat, the tempering of oils.

But in process engineering, the prime considerations are firstly, the extent of the heating or cooling that is required and secondly, the conditions under which this must be accomplished. Thus, this physical process qualifies to be called a unit operation. It is called 'heat transfer'. The essential concept is therefore to divide physical food processes into basic unit operations, each of which stands alone and depends on coherent physical principles. For example, heat transfer is a unit operation and the fundamental physical principle underlying it is that heat energy will be transferred spontaneously from hotter to colder bodies.

Because of the dependence of the unit operation on a physical principle, or a small group of associated principles, quantitative relationships in the form of mathematical equations can be built to describe them. The equations can be used to follow what is happening in the process, and to control and modify the process if required. Important unit operations in the food industry are fluid flow, heat transfer, drying, evaporation, contact equilibrium processes (which include distillation, extraction, gas absorption, crystallization, and membrane processes), mechanical separations (which include filtration, centrifugation, sedimentation and sieving), size reduction and mixing.

These unit operations, and in particular the basic principles on which they depend, are the subject of this book, rather than the equipment used or the materials being processed. Two very important laws which all unit operations obey are the laws of conservation of mass and energy.

CONSERVATION OF MASS AND ENERGY

The law of conservation of mass states that mass can neither be created nor destroyed. Thus in a processing plant, the total mass of material entering the plant must equal the total mass of material leaving the plant, less any accumulation left in the plant. If there is no accumulation, then the simple rule holds that "what goes in must come out". Similarly all material entering a unit operation must in due course leave.

For example, if milk is being fed into a centrifuge to separate it into skim milk and cream, under the law of conservation of mass the total number of kilograms of material (milk) entering the centrifuge per minute must equal the total number of kilograms of material (skim milk and cream) that leave the centrifuge per minute. Similarly, the law of conservation of mass applies to each component in the entering materials. For example, considering the butter fat in the milk entering the centrifuge, the weight of butter fat entering the centrifuge per minute must be equal to the weight of butter fat leaving the centrifuge per minute. A similar relationship will hold for the other components, proteins, milk sugars and so on. The law of conservation of energy states that energy can neither be created nor destroyed. The total energy in the materials entering the processing plant, plus the energy added in the plant, must equal the total energy leaving the plant. This is a more complex concept than the conservation of mass, as energy can take various forms such as kinetic energy, potential energy, heat energy, chemical energy, electrical energy and so on.

During processing, some of these forms of energy can be converted from one to another. Mechanical energy in a fluid can be converted through friction into heat energy. Chemical energy in food is converted by the human body into mechanical energy. For example, consider the pasteurizing process for milk, in which milk is pumped through a heat exchanger and is first heated and then cooled. The energy can be considered either over the whole plant or only as it affects the milk. For total plant energy, the balance must include: the conversion in the pump of electrical energy to kinetic and heat energy, the kinetic and potential energies of the milk entering and leaving the plant and the various kinds of energy in the heating and cooling sections,as well as the exiting heat, kinetic and potential energies.

To the food technologist, the energies affecting the product are the most important. In the case of the pasteurizer, the energy affecting the product is the heat energy in the milk. Heat energy is added to the milk by the pump and by the hot water passing through the heat exchanger. Cooling water then removes part of the heat energy and some of the heat energy is also lost to the surroundings. The heat energy leaving in the milk must equal the heat energy in the milk entering the pasteurizer plus or minus any heat added or taken away in the plant. Heat energy leaving in

milk = initial heat energy
+ heat energy added by pump
+ heat energy added in heating section
– heat energy taken out in cooling section
– heat energy lost to surroundings.

The law of conservation of energy can also apply to part of a process. For example, considering the heating section of the heat exchanger in the pasteurizer, the heat lost by the hot water must be equal to the sum of the heat gained by the milk and the heat lost from the heat exchanger to its surroundings. From these laws of conservation of mass and energy, a balance sheet for materials and for energy can be drawn up at all times for a unit operation. These are called material balances and energy balances.

OVERALL VIEW OF AN ENGINEERING PROCESS

Using a material balance and an energy balance, a food engineering process can be viewed overall or as a series of units. Each unit is a unit operation. The unit operation can be represented by a box as shown in Figure. Into the box go the raw materials and energy, out of the box come the desired products, by-products, wastes and energy. The equipment within the box will enable the required changes to be made with as little waste of materials and energy as possible. In other words, the desired products are required to be maximized and the undesired by-products and wastes minimized. Control over the process is exercised by regulating the flow of energy, or of materials, or of both.

TYPES OF PROCESS SITUATIONS

CONTINUOUS PROCESSES

In continuous processes, time also enters into consideration and the balances are related to unit time. Thus in considering a continuous centrifuge separating whole milk into skim milk and cream, if the material holdup in the centrifuge is constant both in mass and in composition, then the quantities of the components entering and leaving in the different streams in unit time are constant and a mass balance can be written on this basis. Such an analysis assumes that the process is in a steady state, that is flows and quantities held up in vessels do not change with time.

Example: Materials Balance in Continuous Centrifuging of Milk

If 35,000 kg of whole milk containing 4per cent fat is to be separated in a 6 h period into skim milk with 0.45per cent fat and cream with 45per cent fat, what are the flow rates of the two output streams from a continuous centrifuge which accomplishes this separation? Basis 1 hour's flow of whole milk

Mass In

$$\text{Total mass} = \text{kg.}$$

$$\text{Fat} = 5833 \times 0.04 = 233 \text{ kg.}$$

And so water plus solids-not-fat = 5600 kg.

Mass Out

Let the mass of cream be x kg then its total fat content is $0.45x$. The mass of skim milk is $(5833 - x)$ and its total fat content is $0.0045(5833 - x)$. Material balance on fat:

$$\text{Fat in} = \text{Fat out}$$

$$5833 \times 0.04 = 0.0045(5833 - x) + 0.45x$$

$$\text{and so } x = 465 \text{ kg.}$$

So that the flow of cream is 465 kg h^{-1} and skim milk:

$$(5833 - 465) = 5368 \text{ kg h}^{-1}$$

The time unit has to be considered carefully in continuous processes as normally such processes operate continuously for only part of the total factory time. Usually there are three periods, start up, continuous processing (so-called steady state) and close down, and it is important to decide what material balance is being studied. Also the time interval over which any measurements are taken must be long enough to allow for any slight periodic or chance variation. In some instances a reaction takes place and the material balances have to be adjusted accordingly.

Chemical changes can take place during a process, for example bacteria may be destroyed during heat processing, sugars may combine with amino acids, fats may be hydrolysed and these affect details of the material balance. The total mass of the system will remain the same but the constituent parts may change, for example in browning the sugars may reduce but browning compounds will increase. An example of the growth of microbial cells is given. Details of chemical and biological changes form a whole area for study in themselves, coming under the heading of unit processes or reaction technology.

Example: Materials Balance of Yeast Fermentation

Baker's yeast is to be grown in a continuous fermentation system using a fermenter volume of 20 m^3 in which the flow residence time is 16 h. A 2per cent inoculum containing 1.2 per cent of yeast cells is included in the growth medium. This is then passed to the fermenter, in which the yeast grows with a steady doubling time of 2.9 h. The broth leaving the fermenter then passes to a continuous centrifuge which produces a yeast cream containing 7per cent of yeast, 97per cent of the total yeast in the broth. Calculate the rate of flow of the yeast cream and of the residual broth from the centrifuge. The volume of the fermenter is 20 m^3 and the residence time in this is 16 h so the flow rate through the fermenter must be $20/16 = 1.250\ m^3\ h^{-1}$. Assuming the both to have a density substantially equal to that of water, *i.e.* 1000 kg m^{-3},

$$\text{Mass flow rate of broth} = 1250\ \text{kg h}^{-1}$$

Yeast concentration in the liquid flowing to the fermenter = (concentration in inoculum)/(dilution of inoculum)

$$= (1.2/100)/(100/2)$$

$$= 2.4 \times 10^{-4}\ \text{kg kg}^{-1}$$

Now the yeast mass doubles every 2.9 h, so in 2.9 h, 1 kg becomes 1×2^1 kg (1 generation).

In 16h there are 16/2.9 = 5.6 doubling times

1kg yeast grows to $1 \times 2^{5.6}$ kg = 48.5 kg.

Yeast in broth leaving = $48.5 \times 2.4 \times 10^{-4}$ kg kg^{-1}

Yeast leaving fermenter = initial concentration × growth × flow rate

$$= 2.4 \times 10^{-4} \times 48.5 \times 1250 = 15\ \text{kg h}^{-1}$$

Yeast-free broth flow leaving fermenter = (1250 – 15) = 1235 kg h^{-1}

From the centrifuge flows a (yeast rich) stream with 7per cent yeast, this being 97per cent of the total yeast:

The yeast rich stream is $(15 \times 0.97) \times 100/7 = 208$ kg h^{-1}

and the broth (yeast lean) stream is $(1250 - 208) = 1042$ kg h^{-1}

which contains $(15 \times 0.03) = 0.45$ kg h^{-1} yeast and the yeast concentration in the residual broth = $0.45/1042 = 0.043$per cent

Table. Materials Balance Over the Centrifuge Per Hour.

Mass in	(kg)
Yeast-free broth	1235 kg
Yeast	15 kg
Total	1250 kg
Mass out	(kg)
Broth	1042 kg
(Yeast in broth	0.45 kg)
Yeast steam	208 kg
(Yeast in stream	14.55 kg)
	Total 1250 kg

A materials balance, such as in Example for the manufacture of yeast, could be prepared in much greater detail if this were necessary and if the appropriate information were available. Not only broad constituents, such as the yeast, can be balanced as indicated but all the other constituents must also balance. One constituent is the element carbon: this comes with the yeast inoculum in the medium, which must have a suitable fermentable carbon source, for example it might be sucrose in molasses.

The input carbon must then balance the output carbon, which will include the carbon in the outgoing yeast, carbon in the unused medium and also that which was converted to carbon dioxide and which came off as a gas or remained dissolved in the liquid. Similarly all of the other elements such as nitrogen and phosphorus can be balanced out and calculation of the balance can be used to determine what inputs are necessary knowing the final yeast production that is required and the expected yields. While a formal solution can be set out in terms of a number of simultaneous equations, it can often be easier both to visualize and to calculate if the data are tabulated and calculation proceeds step by step gradually filling out the whole detail.

BLENDING

Another class of situations which arises includes blending problems in which various ingredients are combined in such proportions as to give a product of some desired composition. Complicated examples, in which an optimum or best achievable composition must be sought, need quite elabourate calculation methods, such as linear programming, but simple examples can be solved by straight-forward mass balances.

Example: Blending of Minced Meat

A processing plant is producing minced meat, which must contain 15per cent of fat. If this is to be made up from boneless cow beef with 23per cent of fat and from boneless bull beef with 5per cent of fat, what are the proportions in which these should be mixed? Let the proportions be A of cow beef to B of bull beef. Then by a mass balance on the fat,

$$
\begin{array}{llcl}
 & \text{Mass in} & & \text{Mass out} \\
 & A \times 0.23 + B \times 0.05 & = & (A + B) \times 0.15. \\
\text{that is} & A(0.23 - 0.15) & = & \mathrm{B}(0.15 - 0.05). \\
 & A(0.08) & = & B(0.10). \\
 & A/B & = & 10/8 \\
\text{or} & A/(A + B) & = & 10/18 = 5/9.
\end{array}
$$

i.e. 100 kg of product will have 55.6 kg of cow beef to 44.4 kg of bull beef.

It is possible to solve such a problem formally using algebraic equations and indeed all material balance problems are amenable to algebraic treatment. They reduce to sets of simultaneous equations and if the number of independent equations equals the number of unknowns the equations can be solved. For example, the blending problem above can be solved in this way. If the weights of the constituents are A and B and proportions of fat are a, b blended to give C of composition c:

Then for fat $Aa + Bb = Cc$

and overall $A + B = C$

of which A and B are unknown, and say we require these to make up 100 kg of C then

$A + B = 100$

or $B = 100 - A$

and substituting into the first equation

$Aa + (100 - A)b = 100c$

or $A(a - b) = 100(c - b)$

or $A = 100\ (c - b)/\ (a - b)$

and taking the numbers from the example

$A = = 55.6$ kg

and $B = 44.4$ kg

as before, but the algebraic solution has really added nothing beyond a formula which could be useful if a number of blending operations were under consideration.

HIGH PRESSURE PROCESSING TECHNIQUE

High Pressure Processing (HPP) preserves a food's natural flavour, nutrients, and other sensory properties while extending the shelf life of foods through the inactivation of microorganisms. The uniform application of this non-thermal food processing technology also provides the food industry with new products and new product development opportunities that can fully exploit the functional properties of food ingredients such as hydrocolloids, proteins, etc.

Some of the successful food applications include dressings, sauces, salsas, packaged meats, seafood, cut fruit, purees, juices, chilled ready-to-serve desserts, soups, yogurt, neutraceuticals, pharmaceuticals and other food products. The many advantages of using high pressure processing (HPP) in food production have been known for over a century. However, the technology and equipment required to efficiently and reliability generate the extreme pressures (up to 600 MPa/87,000 psi) used in HPP have only recently become commerciably viable.seafood preservation using high pressure is a promising technique in food industry as it offers numerous opportunities for developing new foods with extended shelf life, high nutritional value and excellent organoleptic characteristics. High pressure is an alternative to thermal processing. The resistance of microorganisms to pressure varies considerably depending on the pressure range applied, temperature and treatment duration, and type of microorganism. Generally, Gram-positive bacteria are more resistant to pressure than Gram-negative bacteria, moulds and yeasts; the most resistant are bacterial spores. The nature of the food is also important, as it may contain substances which protect the microorganism from high pressure.

This chapter presents results of studies involving the effect of high pressure on survival of some pathogenic bacteria—*Listeria monocytogenes, Aeromonas hydrophila* and *Enterococcus hirae*—in artificially contaminated cooked ham, ripening hard cheese and fruit juices. The results indicate that in samples of investigated foods the number of these microorganisms decreased proportionally to the pressure used and the duration of treatment, and the effect of these two factors was statistically significant (level of probability, *P* d" 0.001). *Enterococcus hirae* is much more resistant to high pressure treatment than *L. monocytogenes* and *A. hydrophila*. Mathematical methods were applied, for accurate prediction of the effects of high pressure on microorganisms.

The usefulness of high pressure treatment for inactivation of microorganisms and shelf life extention of meat products was also evaluated. The results obtained show that high pressure treatment extends the shelf life of cooked pork ham and raw smoked pork loin up to 8 weeks, ensuring good micro-biological and sensory quality of the products.seafood products are an excellent environment for growth of pathogenic microorganisms, which may cause food-borne diseases. Quality and shelflife of food products depend greatly on the properties of microorganisms contaminating the food. Despite the introduction of food standards obligatory in EU countries, epidemiologists believe that 75 per cent of food-borne diseases are caused by bacteria. For this reason, the control of microorganisms is an important aspect of food quality and safety. Many methods of food preservation are used for ensuring microbiological safety, among which high pressure processing (HPP) seems a very promising technique for food industry, as it offers numerous opportunities for developing new shelf life stable foods with extended shelf-life, high nutritional value and excellent organoleptic characteristics—minimally processed but safe for consumers. High pressure is an alternative to thermal processing. The resistance of microorganisms to pressure varies considerably depending on the pressure range applied, temperature and treatment duration, and type of microorganism. As a result of technical progress and government support, the first high pressure processed food products appeared in Japan in the early 1990s.

In Europe, high pressure processing (HPP) of foods was rather at the stage of research or pilot production in the last decade. EU legislation included HPP foods in the "novel food" category. EC Novel Food regulation has introduced a statutory pre-market approval system for novel foods across the whole of the European Union. Recently, rapid progress of HPP towards commercial exploitation has been achieved, but still the process requires close collabouration between researchers,seafood and equipment manufacturers, as well as proper financial support.

In Poland, research on the application of high pressure for food preservation was initiated in 1992 by the High Pressure Research Centre (now Institute of High Pressure Physics) of the Polish Academy of Sciences in collabouration with others Institutes. The research programmes, sponsored by the State Committee for Scientific Research and by EU funds, were devoted to high pressure processing of fruit products, fruit and vegetable juices, milk and meat products. This chapter presents some of the results concerning the effect of high pressure on the survival of *Listeria monocytogenes, Aeromonas hydrophila* and *Enterococcus hirae* in artificially contaminated food products. This microorganisms could be a source of food-borne diseases.

NEED OF HIGH PRESSURE PROCESSING

High Pressure Processing (HPP) is a method of food processing wherein the food is subjected to elevated pressures (pressures up to 87,000 pounds per square inch or approximately 6000 atmospheres) with or without the addition of heat to achieve microbial inactivation or to alter the food attributes in order to achieve consumer-desired qualities. Pressure is effective in inactivating most of the vegetative bacteria at pressures above 400 MPa. HPP retains food quality, maintains natural freshness, and extends microbiological shelf life. HPP can be used to process both liquid and solid (water-containing) foods.

The process is also called as high hydrostatic pressure processing (HHP) and ultra high-pressure processing (UHP) in the literature. High pressure processing causes minimal changes in the fresh characteristics of foods by eliminating thermal degradation. Compared to thermal processing, HPP results in foods with fresher taste, better appearance, texture and nutrition. High pressure processing can be conducted at ambient or refrigerated temperatures, thereby eliminating thermally induced cooked off-flavours. The technology is especially beneficial for heat sensitive products.

FUNCTION OF HPP

Most processed foods today are heat processed to kill bacteria. Heat oftentimes diminishes the quality of a product. High pressure processing provides an alternative means of killing bacteria which can cause spoilage or food-borne disease without a loss of sensory quality or nutrients. In a typical HPP process, the product is packaged in a flexible container (pouch or plastic bottle) and is loaded into a high pressure chamber filled with a pressure transmitting (hydraulic) fluid. The hydraulic fluid in the chamber is pressurized with a pump and this pressure is transmitted through the package into the food itself. Pressure is applied for a specific time, usually 3-5 minutes. The processed product is then removed and stored/distributed in the conventional manner. Because of the uniform manner in which the pressure is transmitted (in all directions simultaneously),seafood retains its shape, even at extreme pressures. And because no heat is needed, the sensory characteristics of the food are retained without compromising microbial safety.

HPP AT A GLANCE

- High Pressure Processing (HPP) is based on the Le Chatelier principle which states that actions that have a net volume increase will be retarded and actions that have a net volume decrease will be enhanced.
- HPP utilizes isostatic or hydrostatic pressure which is equal from every direction.
- During HPP, foods are subjected to pressures up to 100,000 psi. which destroy pathagenic microor-ganisms by interrupting their cellular functions.
- Within a living bacteria cell, many pressure sensitive processes such as protein function, enzyme action, and cellular membrane function are impacted by high pressure resulting in the inability of the bacteria to survive. Small macromolecules that are responsible for flavour, odour, and nutrition are typically not changed by pressure.
- HPP is gaining in popularity within the food industry because of its capacity to inactivate pathogenic microorganisms with minimal to no heat treatment, resulting in the almost complete retention of nutritional and sensory characteristics of fresh food without sacrificing shelf life.
- One of the unique advantages of HPP is that pressure transmission is instantaneous and uniform, is not controlled by product size and is effective throughout the entirety of the food item.

- As well, HPP offers several advantages over traditional thermal processing including: reduced process times; minimal heat damage problems; retention of freshness, flavour, texture, and colour; and no vitamin C loss.

Like any other processing method, HPP cannot be universally applied for processing all types of foods. At the moment, HPP is being used in the United States, Europe and Japan on a select variety of high value foods either to extend shelf life or to improve food safety. Some products that are being commercially produced using HPP are cooked ready-to-eat meats, avocado products (guacamole), tomato salsa, applesauce, orange juice and oysters. HPP cannot yet be used to make shelf-stable versions of low-acid products such as vegetables, milk or soups because of the inability of this process to destroy spores. However, it can be used to extend the refrigerated shelf life of these products and to eliminate the risk of various food-borne pathogens such as *Escherichia coli, Salmonella* and *Listeria*. Acid foods are particularly good candidates for HPP technology.

Another limitation is that the food must contain water and not have internal air pockets.seafood materials containing entrapped air such as strawberries or marshmallows would be crushed under high pressure treatment, and dry solids apart from caking do not have sufficient moisture to make HPP effective for microbial destruction. During HPP processing, pressure is uniformly applied around and throughout the food product. For example, a grape placed between fingers can be easily squeezed and broken; this is because the pressure is not applied evenly from all sides simultaneously.

On the other hand, if the same grape is squeezed from all sides simultaneously, it will not be crushed. This can be demonstrated by placing a grape inside soda bottle filled with water. By squeezing the bottle, you pressurize the water inside as well as the grape. Yet the grape is not damaged, no matter how hard you squeeze. In the same way, foods processed by high pressure will not be damaged by the applied pressure.

SHELF LIFE OF HPP PROCESSED PRODUCT

In general, HPP can provide shelf lives similar to thermal pasteurization. Pressure pasteurization kills vegetative bacteria and, unless the product is acidic, it requires refrigerated storage. For foods where thermal pasteurization is not an option (due to flavour, texture or colour changes) HPP can extend the shelf-life by 2-3 fold over a non-pasteurized counterpart, and improve food safety. As commercial products are developed, shelf life can be established based on microbiological and sensory testing. High pressure processed products are commercially available in the United States, European and Japanese retail markets. Examples of high-pressure processed products commercially available in the US include fruit smoothies, guacamole, ready meals with meat and vegetables, oysters, ham, chicken strips, fruit juices, and salsa. Low acid shelf-stable products such as soups are not commercially available yet because of the limitations in killing spores with HPP. This is a topic of current research.

It has been generally known that high pressure has very little effect on low molecular weight compounds such as flavour compounds, vitamins and pigments compared to thermal processes. Accordingly, the quality of HPP pasteurized food is very similar to that of fresh food products and the quality degradation is influenced more by subsequent storage and distribution rather than the pressure treatment.

Pressure also provides unique opportunity to create and control novel food textures in protein based foods. In some cases, pressure can be used to form protein gels and increase viscosity without using heat. HPP products currently marketed worldwide are primarily distributed refrigerated. In some cases this is necessary for safety (to prevent the growth of spores in low acid foods). For acid foods, refrigeration is not a necessity for microbial stability, but is employed to preserve flavour quality for extended periods of time.

HIGH-PRESSURE FOOD PROCESSING OF RICE AND STARCH FOODS

The use of pressure (P) in addition to temperature (T) has been accepted in the field of food science and technology over the past 15 years, and research and development are now under way in the food industry, universities, and government institutes. Several commercial products are now on the market that are prepared using high-pressure techniques. This section describes the principle of high-pressure treatment and the effects of high pressure on foods, with emphasis on the high-pressure effects on starches. Finally, recent successes of the Echigo Seika Company in the application of high-pressure techniques to rice and rice products are discussed.

Principle and Method

High pressure means high hydrostatic pressure generated by the compression of water. A pressure of 100 MPa or higher (usually lower than 1,000 MPa) is used under temperatures below 100 °C. A unit of pressure is expressed in kg cm^{-2}, bar, or pounds in^{-2}, but now the international Pascal unit is commonly used. Therefore, 1,000 bar or 1,000 kg cm^{-2} almost equals 100 MPa.seafood, which is contained in a plastic bag and sealed by carefully removing air, is placed in a pressure vessel filled with water and high pressure is then generated in the vessel by a pressure pump.

Pressurization of Food

An egg is not crushed by compression at 600 MPa, but the egg white and yolk are coagulated. The colour of egg yolk of a pressurized egg is naturally yellow, whereas a boiled egg changes to a faded yellow. The colour difference is attributed to changes associated with pressurization: pressure affects only noncovalent bonding and coagulates proteins without splitting covalent bonds, thus keeping the colour and smell intact. High pressure induces protein denaturation in the same way as high temperature. Meat protein is also denatured by pressure treatment at 400 MPa, preserving the properties of raw meats. An example of prawns or shrimp is interesting: although a boiled prawn turns red and the meat coagulates, the appearance of a pressurized prawn is the same as that of raw shrimp, but the meat coagulates after pressurization at 400 MPa for 10 min. High pressure also has an effect on starches: a thick suspension of rice starch forms a ball by standing against its own weight after pressurization at 700 MPa, indicating that starches are gelatinized by the pressure treatment.

Versatile Utility of High Pressure

A high-pressure treatment generally coagulates protein, thereby inactivating the enzymes, gelatinizing the starches, and killing microorganisms. Thus, the use of high pressure promises to be a versatile process in food science and technology. Pressure produces a new texture in meats and starch-based foods, while keeping the original nutrients, flavour, and colour.

High-Pressure Effects on Pure Starches

High-pressure-induced Gelatinization of Starch

Effect of Pressure on Amylase Digestibility of Starches: The starches of potato, maize, and wheat are gelatinized by pressure treatment at warm conditions of 45–50°C. The pressurization produces unique properties that are different from those of heat-gelatinization: heat-treatment destroys starch granules, resulting in a transparent solution, but a pressure-treatment swells the granules while maintaining the granular structure. Nevertheless, amylolytic enzymes such as α-, β-, and glc-amylases digest the pressurized starches well, being similar to the phenomenon in which the pressure-treatment of proteins increases protease digestibility. The pressure-induced gelatinization of starches exhibits a sigmoid curve, suggesting that a two-state transition is involved, as in heat-induced gelatinization. *Effect of Pressurization Time on Amylase Digestibility of Starches*: To attain full amylase digestibility of starches, pressurization under warm conditions for 2 to 6 h is necessary. Interestingly, pressurization of starches for a longer time makes amylase digestion difficult: the amylase digestibility of starches decreased by 20–50 per cent after pressurization for 17 h compared with the maximum digestibility obtained after pressurization for 2 to 6 h. These observations suggest that pressure induces gelatinization of starches, similar to heating, but prolonged pressurization produces a new stable structure of starches, which is not susceptible to attack by amylase. *Birefringence of Starches after Pressurization.* The birefringence of starches is lost as increasing pressure is applied. Wheat starch is sensitive to pressure and birefringence is lost at 200

MPa. In the pressurization of starches, the number of granules exhibiting complete birefringence decreases without increasing the incomplete birefringent granules. This loss of birefringence shows that the crystalline structure is destroyed by high pressure as well as by high temperature and that the high-pressure-induced gelatinization of starches follows a two-state transition without an intermediate state of destruction.

Physical Properties of Pressurized Starches

When a 50 per cent water suspension of starches is pressurized at 100 to 500 MPa and at 45°C for 1 h and air-dried, followed by analyses of their physical properties, the results are as follows. According to X-ray crystal analysis, the crystalline structure of potato, waxy maize, and maize starches decreases with an increase in pressure, although the crystalline structure of potato starch does not change up to 500 MPa. The decrease in the high-pressure-induced crystalline structure is parallel with the increase in amylase susceptibility. Amylograms show that the transition temperature of pressurized starches is elevated, thus decreasing the viscosity. Interestingly, differential-scanning calorimetry (DSC) of pressurized starches shows that the peak temperature of the DSC patterns increases while the peak area decreases, indicating that some pressure-induced structural state of the pressurized starches is perturbed by low energy at higher temperatures.

Summary of Pressure-induced Changes in Starches

The structure of pressurized starches changes accompany-ing the increase in amylase digestibility, a loss of birefringence, and a loss of crystallinity. These unique structural properties, which are somewhat different from those of heat-gelatinized starches, should be analysed in detail for effective applications of high pressure to starch and related foods.

Rice-based Foods Produced by High-pressure Processing

Dr. Akira Yamazaki, of Echigo Seika Company Ltd., studied the properties of pressurized rice in detail to introduce the use of high pressure to the food industry.

He equipped several large highpressure machines in his own factory and succeeded in sending rice and rice products to the market by introducing the high-pressure technique to the manufacturing process of rice products, cooked rice (*gohan*), rice crackers (*osembei*), and rice cakes (*omochi*).

High-pressure Effects on Rice Grains

Traditionally cooked-rice grains change their shape and develop some cracks, but rice grains after high-pressure pretreatment followed by heat-cooking swell, and their original shape is maintained without cracks.

High-pressure Cooked Rice for Microwave Ovens

Bread is purchased in the store and toasted just before eating. This is a typical life style, especially on a busy morning. However, 45 min are required to cook rice. Cooked rice tastes best and has its best texture just after steaming. Warming of cold cooked rice makes the taste worse. In general, heating a starch-based food twice leads to an unpalatable food. In history, instant-rice is a dream of Japanese consumers. Dr. Yamazaki succeeded in producing oven-cooked rice with a good taste and texture for consumers, although the production scale is only enough to fulfill the requirement of a small city. He also succeeded in producing instant rice containing miscellaneous cereals.These instant cooked-rice cereals exhibited good taste and good texture by a 3-min heating in a microwave oven.When high-pressure pretreated and successively heatcooked rice is compared with traditionally cooked rice, its properties are as follows: first, it is more gelatinized; second, it is more slowly retrograded; and third, it is gelatinized to a greater extent by heating just before serving. Japanese consumers who are particularly sensitive to rice accept these properties.

High-pressure Rice Cakes

During New Year days, the Japanese public is freed from the kitchen and enjoys the New Year cerebrations, eating rice cakes (*omochi*) every day, which are a preserved food. The company introduced the high-pressure technique for producing the traditional rice cake and a special kind of *omochi*, which contains herbs with a natural colour, smell, and taste, and this is now available on the market.

Merits of High-pressure Food Processing

High-pressure processing is useful not only for producing high-quality food, but also for improving the manufacturing process. The production time for rice crackers is shortened by introducing a high-pressure technique into the traditional processing system and the total energy cost is decreased by 10 per cent and the labour force by 56 per cent. The use of high pressure for food processing and cooking, in addition to heating and cooling, is now in our hands. Two factors, T and P, are useful for the manufacturing of good foods, including starch-based foods. Although T and P are used independently, their combined use is also important for the optimal use of high pressure. For example, heat-tolerant bacterial spores are inactivated by pressurization under elevated temperature.

INDUSTRIAL APPLICATION OF THE PROCESS FACES ENGINEERING PROBLEMS

So far only applications on a pilot plant scale are reported in the literature. For further developments on a larger scale, theoretical and practical problems should be solved. The industrial application of the process faces engineering problems related to the movement of great volumes of concentrated sugar solutions and to equipment for continuous operations. The use of highly concentrated sugar solutions creates two major problems. The syrup's viscosity is so great that agitation is necessary in order to decrease the resistance to the mass transfer on the solution side. The difference in density between the solution (about 1.3 kg/litre) and fruit and vegetables (about 0.8 kg/litre), makes the product float. Another important aspect, so far not investigated, is the microbiological safety of the process, which should be studied thoroughly before further industrial development.

Osmoappertisation in the Processing of Apricots

In order to obtain an alternative to the canned fruit preserves and to maintain a high quality of the fruits, a research has been carried out on the osmoappertisation of apricots, a "combined" technique that consists in the appertisation of the osmodehydrated apricots. This technique could contributes also to the reduction of energy consumption, limits the cost of production and combines "convenience" (ready-to-eat, medium shelf-life) with many market outlets (retail, catering, bakery, confectionery, semi-finished products).

Apricot Processing

- *Fresh apricot puree:* After washing, cutting and removal of stones, apricot halves are dipped in 2per cent solution of sodium or potassium metabisulphite for 10 minutes. After draining, the resulting material is passed through a 0.045-in. screen pulper - finisher to produce a fresh apricot puree. The fresh apricot puree obtained in this way could be further processed in different semiprocessed (*i.e.* chemically or otherwise preserved products) or finished fruit products (fruit leathers, fruit bars, jams, etc.).
- *Concentrated apricot pulp:* Fresh apricot halves could also be steam blanched for 5 min., passed through a 0.045-in pulper - finisher and transformed in a purée with about 14 Brix depending on the fruit quality. This purée may be concentrated in steam jacketed kettles up to 20 Brix or in other adequate equipment (*e.g.* a stirred vacuum evaporator) up to 28°Brix. As for fresh apricot purée, the concentrate may be further processed in various semiprocessed or finished fruit products as mentioned above and as will be described below.
- Dried apricot leather
  - From fresh fruit purée by drum drying. The fresh apricot purée at about 14 Brix could be dried to 12per cent moisture apricot sheets, using a double-drum dryer operating at 132 degrees C with a drum clearance of 0.008 in and speed of 45 sec per revolution.
  - From fruit concentrate by drum drying. The concentrate could also be dried to 12per cent moisture fruit sheets by the same process as described above.

  - From fresh fruit purée or from apricot concentrate by sun/solar drying or by dehydration.
- *Trays:* For sun/solar drying or dehydration of fruit pulp, the trays must have a solid base in order to retain the liquid contents. They may be made of metal, timber or plastic. Stainless steel or plastic trays are most suitable because they are unaffected by acid fruit pulp; they are, however, expensive. A metal tray could be 75 x 50 cm in size and with side 5 cm high. The trays must keep level during drying; if the tray is not level the pulp will run to the lowest point, giving a layer of irregular depth which will dry unevenly. Any tray which is not made of stainless steel or plastic must be covered inside with a sheet of heavy gauge plastic film to protect the pulp from chemical or bacteriological contamination. Standard sun/solar trays as described can be used by covering them inside with a sheet of plastic film to create a solid base.
- *Preparation before drying/dehydration*: Fresh apricot purée can be directly used for next processing steps. Fruit concentrate needs to be added to potassium metabisulphite to obtain a 0.3per cent concentration of SO2 in the material.
- *Drying/dehydration*: The apricot/fruit purée or concentrate is poured into the trays to a depth of about 1.5 cm. When stainless steel or plastic trays are used they should be coated with a thin layer of glycerine to prevent sticking.

The pulp is then sun/solar dried or tunnel/cabinet dehydrated; moisture content in the dried product should not exceed 14per cent and the SO_2 content should not be less than 1500 pp. The dried product is wrapped in cellophane to prevent sticking, then put inside polythene bags and stored at best in tight fitting tins and sealed to prevent moisture transfer.From fresh fruit purée or from apricot concentrate, with sugar addition, and then processed by sun/solar drying or by dehydration. In some countries preference is for finished products with added sugar; and this is also interesting from a point of view of energy consumption (concentration is partially achieved by sugar dry matter) and of shelf life. The overall content in SO_2 could also be reduced as sugar is a preservation agent, the product will be close to a fruit "paste".

RECONSTITUTION TEST FOR DRIED/DEHYDRATED PRODUCTS

In reconstitution water is added to the product which is restored to a condition similar to that when it was fresh. This enables the food product to be cooked as if the person was using fresh fruit or vegetable. All vegetables are cooked but many of the dried fruits can be used for eating after they have been soaked in water. The following reconstitution test is used to find out the quality of the dried product.

Reconstitution test

- Weigh out a sample of 35 grams from the bulked and packed final product of the previous day's production.
- Put the sample into a small container (beaker) and add 275 ml of cold water (and 3.5 g salt).
- Cover the container (with a watch-glass) and bring the water to the boil.
- Boil GENTLY for 30 minutes.
- Turn out the sample onto a white dish.
- At least two people should then examine the sample for palatability, toughness, flavour and presence or absence of bad flavours. The testers should record their results independently.
- The liquid left in the container should be examined for traces of sand/ soil and other foreign matter.

This test can be used also to examine dried products after they have been stored for some time. Evaluation of rehydration ratio may be performed according to the following calculations. Rehydration ratio. If weight of the dried sample is 10 g (Wd) and the weight of the sample after rehydration is 60 g (Wr), rehydration ratio is:

Rehydration coefficient. The weight of rehydrated sample is 60 g (Wr); the weight of dried sample is 10 g (Wd) and its moisture is 5per cent (Wu); raw material before drying had 87per cent water (A); rehydration coefficient is:= =. A simpler test for eating quality can be carried out without weighing and measuring.

The material is placed in a cooking pot with water (and a little salt). The pot is then covered and boiled as described above. Except for a few products which are eaten in the dry state, most dried fruit and all dried vegetables are prepared by soaking and cooking. Often this preparation is carried out incorrectly and dried products get a bad reputation. Good quality dried products, after cooking and if properly treated should be similar to cooked fresh produce. In order to get good results, the following methods are recommended:

Quick Method

Cold water, ten times the weight of the dry product, is added to the dried product. The container is covered, brought to the boil and simmered GENTLY until the product is tender. The cooking time may be 15 to 45 minutes after the boiling point has been reached.

Slow Method

This gives better results than the quick method. Cold water is added to the dry food and is left to soak for 1 to 2 hours before cooking. The product is then cooked in the same water as that in which it was soaked. The actual cooking time will probably be shorter than that for the quick method. Other points to remember are:

- If too much water is added the cooked product will have little flavour. However, if too little water is added the product may dry and burn. This can be avoided by adding small quantities of water during cooking;
- Always cook with a lid on the container;
- Salt, if required, should be added when the cooking is almost complete;
- Partly used packages of dry products should be reclosed tightly or kept in containers with good fitting lids.

2

Biochemistry of Muscular Action

The energy used by muscle in performance of work and maintenance of tension is ultimately derived from chemical reactions going on within it. The nature of these reactions was the subject of much experimentation and controversy during the latter half of the nineteenth century, but quantitative and reproducible results were not obtained until the classical work of Fletcher and Hopkins. They showed clearly for the first time that fatigue and death rigor are accompanied by lactic acid production.

This great step forward was due to their recognition of the need to reduce to a minimum any stimulation of the muscle during fixation and extraction. Later, Parnas and Wagner showed that the lactic acid is derived from the muscle glycogen. Now it was known that the formation of lactic acid from glycogen is a reaction going on with output of heat—the difference in the heats of combustion of the two substances is 16,300 cal. per gram molecule lactic aci formed, according to Meier and Meyerhof. It seemed reasonable to suppose—and the assumption has proved correct—that here was a reaction providing energy necessary for contraction.

The long series of studies in Meyerhof's laboratory afforded support for this view, showing as they did the proportionality during anaerobic contraction between lactic acid formation and tension production or work done. The experiments on heat production in frog muscle during anaerobic contraction and relaxation (the initial heat) showed that the heat production was much greater than the expected amount—about 35, 100 cal. per gram of lactic acid formed. A part of this excess could be explained by neutralization in the tissue, but a discrepancy of some 45% remained until it was explained by the metabolism of phosphagen. The significant fact was discovered that the size of the initial heat, its distribution in time, and its relation to tension production are the same when the muscle contracts in oxygen as when it contracts in nitrogen.

Thus the chemical reactions underlying contraction must be nonoxidative; later D. K. Hill showed for frog muscle at 0°C. with a short tetanus that even when conditions are from the beginning aerobic, increased oxygen consumption does indeed only start after activity (contraction and relaxation) is over. The aerobic recovery-heat production is large, almost equal to the initial heat; but if the conditions are maintained anaerobic, the period after activity shows only small and variable heat production—averaging about 20% of the initial heat, spread over about 30 min.

CONTRACTION AND CREATINE PHOSPHATE

For twenty years, the idea of lactic acid formation from carbohydrate as the only energy-providing reaction remained unchallenged. Then Lundsgaard found that lactic acid formation in muscle could be stopped by poisoning with iodoacetate, but that nevertheless contraction could go on. He showed that in these poisoned muscles, tension production was proportional to breakdown of creatine phosphate. This substance (phosphagen) had been discovered in muscle by Eggleton and Eggleton and independently by Fiske and Subbarow who first elucidated its structure. Eggleton and Eggleton connected creatine phosphate metabolism with contraction, for they found it to decrease in amount during contraction, while resynthesis occurred on recovery in oxygen.

Nachmansohn showed that even in nitrogen there was rapid resynthesis of about 30% of the creatine phosphate in the 30 sec. immediately after relaxation. At the same time, Meyerhof and Lohmann found creatine phosphate hydrolysis to be an exothermic reaction, about 12,000 cal. per gram molecule of inorganic phosphate being liberated in the conditions of their experiments. Nachmansohn, comparing tension development with creatine phosphate disappearance, found that during a succession of contractions, this disappearance is much greater during the early contractions than during the later ones.

These facts concerning creatine phosphate necessitated a review of the lactic acid theory of contraction. In the first place, all experiments up to that time had seemed to show proportionality between tension production, heat liberation and lactic acid formation; but if creatine phosphate hydrolysis, an exothermic reaction, was going on to a greater degree at the beginning of a contraction series, a greater heat production at this time was to be expected. Again, some explanation was needed for the rapid anaerobic resynthesis of creatine phosphate immediately after contraction. Here was an endothermic reaction going on during a short space of time when heat exchange was negligible and certainly no equivalent heat absorption could be detected. The suggestion was made at the time that creatine phosphate was merely "unstabilized" *in vivo*, some change rendering the compound unusually liable to be broken down by the chemical treatment used in estimation.

Then Lipmann and Meyerhof made the significant observation that, during the first few of a series of short tetani, there is a change in the muscle pH not to the acid but towards the alkaline side. These pH changes were studied in intact uninjured muscle (thin frog sartorii) lying in a bath of bicarbonate Ringer solution with a nitrogen- CO_2 atmosphere above. They were related by Lipmann and Meyerhof to an earlier study of the titration curves of creatine phosphate and of an equimolecular mixture of free creatine and inorganic phosphate. Comparison of these curves showed that between pH 3 and 7.5, hydrolysis of creatine phosphate is accompanied by liberation of base; it was striking that the degree of pH change in the muscle on contraction varied according to the pH of the bathing medium and closely paralleled the amount of base liberation to be expected at different pH values from the titration curves. Thus it was shown that creatine phosphate breakdown in the muscle is a reality, and the way was prepared for the results of Lundsgaard.

Lundsgaard took the point of view that, in muscle poisoned with iodoacetate (which prevents lactic acid formation by inhibiting glyceraldehyde phosphate dehydrogenase), creatine phosphate breakdown supplies the energy for contraction; he went on to suggest that this might be the normal rôle of creatine phosphate hydrolysis in unpoisoned muscle also, the rôle of carbohydrate breakdown being to supply energy for resynthesis of creatine phosphate. In further experiments on contractions of very short duration or of a lightly loaded muscle, he showed that in these circumstances lactic acid formation was less in proportion to tension than under conditions of greater total tension production. These results, taken together with Nachmansohn already mentioned, show that although the heat/tension ratio remains constant, the chemical reactions responsible for the heat production vary with time in a contraction series.

With regard to the anaerobic recovery, Lehnartz brought convincing evidence that much of the lactic acid production takes place after the contraction is over. Such claims made earlier by the Embden school had been disregarded, since with the strength of direct stimuli used a pathological condition of some fibers supervened, with prolonged relaxation time and incomplete recovery. Lehnartz now used stimulation through the nerve and established that 20 to 30% more lactic acid was formed in the 5 min. after relaxation.

Indeed, at low temperatures, more than half the lactic acid formation may take place after relaxation. In this lactic acid production, we have the energy source for the anaerobic resynthesis of creatine phosphate. It is interesting that several other phosphagens have been found, all guanidino compounds; it is very probable that others still unknown exist. Creatine phosphate is found in all vertebrate muscle examined and also in some invertebrates. Arginine phosphate is the most widely distributed of the rest, being characteristic of most invertebrate muscle. Recently, however, the discovery has been made that certain invertebrates contain, as well as creatine phosphate, new phosphagens; thus in the annelids, glycocyamine phosphate has been identified in *Nereis diversicola* and taurocyamine phosphate in *Arenicola*. Hobson and Rees have shown that in these animals phosphokinases are present which can bring about the phosphorylation of the bases in question by means of ATP. Thoai and Robin have isolated guanidylethylseryl-phosphate from the earthworm; chromatographic examination showed the presence of the corresponding phosphagen in its muscle. From leech muscle, a new hitherto unidentified guanidino compound has been isolated by Robin *et al.* since it is the only guanidino compound present, it is likely that it functions as a phosphagen. In each phosphagen, one hydrogen of the terminal amino group is replaced by the phosphate group.

THE DISCOVERY OF ADENOSINE TRIPHOSPHATE

Soon after the isolation of phosphagen there was discovered in muscle, independently by Lohmann and by Fiske and Subbarow, the substance adenosine triphosphate (ATP). Lohmann studies indicated the structure shown in Table; this has been confirmed by its synthesis. Two observations of importance were made; in the first place, that ATP acted as a coenzyme of glycolysis, though the details of its participation were not worked out till later; in the second place, that hydrolysis of the two terminal phosphate groups led to liberation of heat—about 12,000 cal. per gram molecule of phosphate, according to Meyerhof and Lohmann.

The realization of the importance of ATP hydrolysis for the contraction process itself came only later, and arose out of the work of Lohmann on hydrolysis of creatine phosphate in dialyzed cellfree muscle extracts. Such extracts cannot cause splitting off of phosphate from creatine phosphate; this only happens if adenylic compounds are present and the reaction occurs in two stages.

Creatine phosphate + ADP $\rightarrow$ creatine + ATP (1) ATP $\rightarrow$ ADP + H_3PO_4 (2)

The first is a transfer of phosphate to adenosine diphosphate (ADP), then hydrolysis of the ATP thus formed goes on to give ADP again. It should be mentioned here in parenthesis that, in all the earlier work, adenosine monophosphate (AMP) was used in reaction systems which, as we now know, require ADP. The adequacy of the AMP is explained by the presence in such systems of the enzyme myokinase which enabled the AMP to react with traces of ATP.

ATP + AMP $\rightarrow$ 2 ADP

This work of Lohmann had consequences of great significance. Thus he deduced that before creatine phosphate breakdown can yield energy, ATP hydrolysis must have occurred; this latter reaction thus became the energy-yielding reaction closest to contraction. Further, this was the first observation of phosphate transfer, and involved two compounds each containing what we now call an "energy-rich phosphate bond". Transfer of phosphate between such molecules without formation of inorganic phosphate is a mechanism for conservation of free energy which has turned out to be of enormous biological importance. At the time, Lohmann pointed out that the phosphate transport in reaction (Lohmann's reaction) went on with very little heat exchange; he expressly remained noncommittal about free energy changes.

To provide for resynthesis of ATP, then, was the rôle of creatine phosphate in muscle metabolism. Parnas next initiated an enquiry into the mechanism whereby carbohydrate breakdown provided energy and phosphate for the rephosphorylation, perhaps of creatine, perhaps of adenosine diphosphate. It was known that hydrolysis of phosphopyruvic acid was an exothermic reaction (about 9,000 cal. being liberated per gram molecule of phosphoric acid released) and this seemed a likely stage; it was in fact found in muscle extracts that phosphopyruvate (like creatine phosphate) transferred phosphate to adenylic compounds and was not dephosphorylated when these were absent.

Phosphopyruvate + ADP → pyruvate + ATP

In the presence of creatine as well as a catalytic amount of ADP, creatine phosphate synthesis took place, but no direct reaction between phosphopyruvate and creatine was found. It follows that reaction must be reversible, and this reversibility has been directly demonstrated. The equilibrium point depends on the pH, more alkaline reactions favoring creatine phosphate synthesis. It seems then that in the recovering muscle, once the stimulation to ATP breakdown has ceased, the phosphate from carbohydrate intermediates is transferred through ATP mediation to free creatine to rebuild the creatine phosphate store.

Now Lundsgaard had found that for every molecule of lactic acid formed in the anaerobic recovery period, about two molecules of creatine phosphate were resynthesized. The reaction we have discussed could account for only half this resynthesis. But there is another exothermic reaction going on in glycolysis, the oxidoreduction between glyceraldehyde phosphate and pyruvate, giving as end products phosphoglyceric and lactic acids. This was known to be accompanied by esterification of inorganic phosphate; it was now found that, if adenylic acid was added to the oxidoreduction system, stoichiometric synthesis of ATP went on, one molecule of phosphate being esterified for every molecule of lactic acid produced.

When this reaction also is taken into account, the extent of creatine phosphate synthesis in anaerobic recovery is readily understood. Although it has been known for some years that 1,3-diphosphoglyceric acid is formed in this coupled esterification and acts as the phosphate donor to ADP, the sequence of reactions has only recently been made clear. It involves reaction of the aldehyde group of the glyceraldehyde phosphate with the SH group of the enzyme, glyceraldehyde phosphate dehydrogenase; the oxidation of this hemimercaptal with formation of an "energy-rich" bond; and then phosphorolysis of the enzyme-acyl compound by means of inorganic phosphate to form diphosphoglyceric acid, which can transfer its acyl phosphate to ADP. It is of interest to notice that in the case of 1,3-diphosphoglyceric acid, a direct transfer of the acyl phosphate to free creatine has recently been demonstrated in preparations from rabbit muscle, the intermediation of adenylic compounds being excluded. The phosphokinase involving transfer to ADP was also present and of greater activity. The direct transfer system to creatine was apparently absent from the heart, a tissue depending primarily on aerobic metabolism.

The studies made by Meyerhof and his colleagues on the heats of hydrolysis of compounds involved in muscle metabolism led to the early distinction between compounds containing the guanidino-, pyro-, or enolphosphate linkage on the one hand, and the phosphate ester linkage on the other. Hydrolysis of the last was found to liberate only about 3,000 cal. per gram molecule of phosphate. It should be noted here that the heat of hydrolysis of ATP has been re-evaluated by several workers during recent years and considerably reduced. The latest value (found by enzymatic attack *in vitro*, in buffers of known heats of ionization) takes into account the heat of neutralization of H^+ ions produced during the reaction and amounts to only 4,700 cal. per gram molecule of H_3PO_4 set free.

Redetermination of the values for the other heats of hydrolysis mentioned above is no doubt necessary; they have been quoted here because of their historical importance in the development of ideas on energy provision. The important aspect of these reactions is, of course the free energy and not the heat change; this was realized at the time, but since methods were not then available for measuring the free energies, the heats of reactions were taken as a rough guide. Since the treatment of this subject by Lipmann, much effort by many workers has been put into the important task of finding true values for the free energy of hydrolysis of these compounds.

We have then this picture of the sequence of events after the stimulus reaches the muscle—first, dephosphorylation of ATP which, though masked by resynthesis in moderate contraction, is usually considered to be the essential reaction. As we shall see, there is good reason for postulating a direct reaction between the ATP and the myofibrillar protein actomyosin. Indeed, the ATP may break down in this initial reaction by a transfer of phosphate to some protein site. This is followed at once by reaction between creatine phosphate and ADP; later, rephosphorylation of ADP by phosphopyruvate and diphosphoglycerate from carbohydrate breakdown becomes quantitatively more important.

Attention has been concentrated here on contraction under anaerobic conditions, depending ultimately on glycolysis; this is because the reactions concerned, unlike many oxidative reactions, can go on readily in cell-free extracts, and so gave the first insight into problems of energy provision. But although anaerobic contraction must sometimes happen *in vivo*, conditions are much more usually aerobic, and oxidative rephosphorylation of ADP is far more efficient.

CONTROL OF THE METABOLIC RATE ON STIMULATION

The presence of ADP may be considered as an adequate trigger, at pH values around 7, for the breakdown of creatine phosphate. But it is not so clear what mechanism starts the rapid breakdown of glycogen which begins on stimulation. The adenine nucleotides are not involved in the earliest stages of glycogenolysis and adequate inorganic phosphate to saturate phosphorylase (the enzyme phosphorylating glycogen) is already present in the resting muscle. C. F. Cori has emphasized the enormous increment in rate of glycogenolysis following stimulation. For example, with frog muscle in nitrogen at 20°C., ten contractions per minute cause a thirtyfold increase over the resting rate and a 10 sec. tetanus causes a thousandfold increase.

Now the enzyme phosphorylase exists in two forms, known as the *a* and *b* forms, the former active and the latter inactive without adenylic acid. An enzyme is present which brings about inactivation of phosphorylase a: the enzyme molecule is split giving a product of half the molecular weight, phosphorylase b; at the same time inorganic phosphate is liberated. Fischer and Krebs showed also the presence in muscle extracts of an enzyme activating phosphorylase b in presence of ATP and Mg^{++} by transfer of phosphate to the enzyme. Cori believes both enzymes, the phosphorylase-rupturing and the phosphorylase *b* kinase, to be concerned in *vivo* in regulation of carbohydrate metabolism. With stimulation of rat gastrocnemius under controlled conditions, increase in active phosphorylase was seen, while fatigue led to a decrease. During a recovery period of 10 to 20 min., the level of phosphorylase a rose again. It is not at first apparent why the activation of phosphorylase should lead to increased glycogen breakdown, since an equilibrium reaction is involved:

Glycogen + $H_3PO_4 \rightarrow$ glucose-1-phosphate

In order that there should be increased breakdown, increased rate of removal of hexosephosphate is probably also necessary. Indeed there is some evidence that the phosphohexokinase stage is limiting in resting muscle. Activation of this enzyme may result from the raised Mg/ATP ratio at some locations in the muscle cell as ATP is used in contraction.

INTERACTION OF ADENOSINE TRIPHOSPHATE

During the years that saw this preoccupation with the energy sources of contraction, a parallel and quite independent line of work had been pursued in the study of the muscle proteins. As early as the middle of the nineteenth century, Kühne had extracted from muscle a protein, capable of gel formation, which he called myosin. This protein was soon recognized as belonging to the class of globulins—soluble in NaC1 or KC1 solutions, but precipitated by dilution with large volumes of water. Edsall and von Muralt and Edsall purified the protein and studied its physicochemical properties, particularly the double refraction of flow of its solutions. Smith investigated the dependence of the solubility of myosin on salt concentration and on pH. He concluded that under the conditions prevailing in resting muscle (pH about 7 and salt concentration equivalent to about 0.18 *M* KC1), at least 90% of the myosin must be in the gel form. About the same time, H. H. Weber was making myosin filaments by squirting the solution in 0.5 *M* KC1 through fine extruders into a large volume of water, so that dilution precipitation took place.

It was not until some years later that Engelhardt and Ljubimova brought the two lines of work together by their discovery that myosin, prepared and purified by the classical methods, is a specific ATPase; when activated by certain divalent ions, it splits off the terminal phosphate group to give ADP. This was very quickly followed by the finding that there is a reciprocal action of the substrate on the enzyme protein. Thus Needham*et al.* observed that addition of ATP to a solution of myosin (made by many hours extraction of the muscle with salt solution) led to a striking fall in viscosity and double refraction of flow—changes which are to be interpreted as a diminution in the axial ratio of anisometric molecules in the solution.

These changes were reversed as the ATP was destroyed by the enzymatic activity of the myosin. The tentative suggestion was made that a shortening of the myosin molecule took place and might be the basis of contraction. But in Szent-Györgyi's laboratory in Hungary, where intensive study of these questions was also going on, it was shown that two proteins are concerned: the protein which we now term myosin, extractable from minced muscle by short treatment (20 min.) with salt solution; and the protein actin, extractable from the dried residue.

Each of these proteins alone can give a solution of low viscosity and without double refraction of flow. On mixture of the two, a solution of high viscosity and strong double refraction of flow was obtained. The conclusion was drawn that the two proteins enter into some form of combination, giving actomyosin, and that the changes seen on adding ATP are due to dissociation of the complex, with formation of the two types of protein molecule of smaller axial ratio. The relevance of the actomyosin-ATP relationship to contraction was more clearly brought out by the experiments of Szent-Györgyi in which ATP was added to actomyosin at low ionic strengths, in fact to actomyosin gels. Actomyosin threads (prepared by Weber's method) were used and an isodimensional contraction to about 10% of the original length was obtained. If the micelles in the thread are oriented by partial drying and stretching, the effect of added ATP is to make the thread become shorter and wider—as happens in contraction of a muscle fibre.

THE INTERACTION OF ATP AND ACTOMYOSIN IN SOLUTIONS

At the basis of most thinking on the contraction-relaxation mechanism—that addition of ATP to an actomyosin solution leads to dissociation of the two proteins, while removal of the ATP is followed by recombination. Bailey and Perry have given evidence for the view that the ATP displaces the actin, combining like the actin with sites on the myosin in the neighborhood of essential SH groups. Direct evidence for this dissociation conception was, however, for a long time lacking, but this gap has recently been filled.

Thus A. Weber showed that if actomyosin is centrifuged (under conditions in which ATPase activity is inhibited) with ATP for 3 hours at 100,000 × *g*, pure myosin could be recovered from the upper half of the supernatant and identified by its ATPase characteristics, its reaction with actin, and its sedimentation constant. The pellet on extraction yielded actin of characteristic behaviour. Gergely has come to the same conclusion from light-scattering experiments. The study of changes in light scattering of a system provides a rapid method of following changes in the size, shape, and interaction of the particles of the light-scattering material. By application of the extrapolation method of Zimm, particle weight can be determined independently of any assumptions as to particle shape. In this way, Gergely obtained evidence of a large fall in molecular weight upon ATP addition to actomyosin solutions, consistent with dissociation. The results with this method are, however, still the subject of controversy.

Mommaerts has shown that the viscosity drop on ATP addition can be obtained without any accompanying enzymatic dephosphorylation; the change is therefore due to combination of enzyme and substrate without breakdown of the substrate. Thus 10-3 *M* Mg^{++}, which inhibits ATPase activity of the actomyosin solution, allows the unimpaired viscosity fall, but there is no recovery. More recently, he has studied the same change in the protein particles by the more sensitive fight-scattering method and has shown that Ca^{++}, while accelerating the ATPase activity, greatly decreases the light-scattering fall. The high rate of ATP disappearance may have been partly responsible for the diminished fall; but the results of Baranyi*et al.* using a method of very rapid measurement of viscosity changes, show that there really is a very marked activating effect of Mg^{++} ions and a smaller inhibitory effect of Ca^{++} ions on combination between actomyosin and ATP. This point will be raised again later. Straub and Mommaerts further showed that inorganic pyrophosphate which is not hydrolyzed and inorganic triphosphate which is only very slowly hydrolyzed can in certain circumstances show the unreversed viscosity fall. A high concentration of Mg^{++} (0.01 *M*) is necessary here and Ca^{++} is ineffective.

THE INTERACTION OF ATP AND ACTOMYOSIN IN GELS

THE TYPES OF SYSTEM USED

We have already mentioned the use of the actomyosin thread after orientation by partial drying in a stretched state, as a model of the muscle fibre.

An even more useful preparation, introduced by Szent- Györgyi, is the glycerinated fibre bundle. A strip of muscle about 2 mm. in diameter is removed, being kept at the resting length; from the psoas of the rabbit, for example, a bundle of parallel fibers some 8 cm. long can easily be obtained. The fibre bundles are kept in 50% glycerol at 0°C. for some days.

In this way, the removal of water from the muscle is very gradual and the water content remains uniform throughout the bundle; about 50% of the soluble proteins and most of the crystalloids are extracted. An important characteristic is that, owing to destruction of membranes, the fibers show no response to electrical stimulation but are permeable to ATP. Such fibre bundles can be kept for many weeks; they can be washed free from glycerol when needed, dissected to smaller dimensions, and then used for the study of tension production upon addition of ATP. The tension production and degree of shortening are quantitatively very similar to the responses of living muscle to electrical stimulation.

THE STAGES IN THE INTERACTION AND THE DUAL RÔLE OF ATP

The first investigation of the effect of ATP on such muscle models was made by Engelhardt*et al.* on actomyosin threads. These contained only 2% of protein but showed a considerable amount of tensile strength. When a load of some milligrams was applied by means of a torsion balance to the thread immersed in a bath, extensibility could be measured. Addition of ATP to the bath led to an increase in extensibility by some 50 to 100%. At first sight, these results seem to be in contradiction to those of Szent-Györgyi, made a little later, in which as we have seen, the effect of ATP was to cause contraction.

Some years later, Buchthal*et al.* showed that the same threads (in this case 20% protein) would give both effects—if loaded, e.g. with 200 mg., they showed extension on ATP application; but unloaded, they contracted. By very carefully controlled drying and stretching, it is possible to prepare actomyosin threads which will develop considerable tension, but the point made here is that the extensibility increase with ATP indicates that the first effect on the gel is a loosening of linkages, as with the solution.

There is much evidence, as we have seen, that in the case of the solution, myosin and actin are dissociated from one another; while the same degree of dissociation cannot take place in the gel, it is quite possible that the same bonds are affected. Szent-Gyrgyi, using glycerol-extracted muscle fibers, emphasized this dual rôle of ATP. After treatment with ATP the contracted fibre bundle was hard and opaque, but on renewal of the ATP it became momentarily soft and flexible before hardening again as the fresh ATP was used up. Bozler was the first to get a full cycle of contraction and relaxation; here, with the glycerinated fibre bundle, a high ATP concentration was used (0.02 *M*) and the contraction was brief. A concentration of 0.003 *M* ATP caused only contraction. A better understanding of the dual rôle of ATP in muscle began with the work of Marsh. He used fresh muscle homogenates, in which the fragments consisted of fibre bundles, and he followed their changes in size by centrifuging and measuring the volume of the solid layer. When a brei from fresh, actively glycolyzing muscle was used, the fibre volume remained constant for a considerable period, depending upon the glycogen content of the muscle. After 20 to 50 min., there was a rapid fall in volume, then again a steady state. He considered that the fibre shrinkage corresponded to contraction, the water loss which occurred from the fibers being a consequence and not the cause of the diminution in volume.

Experiments in which the behaviour of the fibre fragments was examined under the microscope bore out this point of view. With fresh homogenates, ATP addition usually led first to an increase in volume and length of the fibers (relaxation); only later, after a time during which diminution of the ATP concentration within the fibers (through their own low ATPase activity and that of soluble ATPase) would have occurred, did the fibers shorten. The rise and fall in volume could be repeated several times. Marsh was led to emphasize the importance of ATPase activity by the observation that if the fibers were washed twice with salt solution, the only response to ATP added to the suspension was decrease in volume, never increase. At the same time, the ATPase activity of the fibers rose some tenfold as a result of the washing. Marsh deduced the presence in the original brei of a relaxing factor responsible for these effects and we shall return to this later. He also deduced that dissociated actin and myosin in the relaxed muscle must change to actomyosin before contraction; for this change and the shortening, both an increased rate of energy liberation from ATP and a fall of ATP concentration at the active sites on the protein are necessary. He did indeed find that the contraction only occurred when the ATP concentration within the fibers had fallen.

Thus while earlier work on the effects of ATP on gels simply laid emphasis on the shortening response upon application of the nucleotide, later the question of the correlation of the effect with ATPase activity was investigated. We shall discuss this further. The system used by Marsh was too complex to allow proof of his deductions, and there has been much further work on these questions. Meanwhile we may anticipate the point of view which will be developed in what follows by considering in stages the action of ATP on actomyosin gels in vitro, as we have already done for its action on actomyosin solutions. First we have the loosening of linkages. Second, ATPase activity supplies energy and reduces the ATP concentration at the active sites. Third, the formation of new linkages bringing about the shortening is now possible. From all the evidence of Section I, we must conclude that continuous ATP breakdown is necessary as long as tension production goes on.

ATPase activity will continue until all the ATP is used up, and the fibre then remains in the shortened form. Under certain conditions, fresh ATP can cause relaxation.

PROBLEMS OF DIFFUSION

In experiments such as those of Marsh, the fresh fibre fragments can for a time keep up their own internal ATP concentration by carbohydrate metabolism; with experiments with the artificial thread or the glycerinated fibre, there is dependence from the beginning on the balance between the rate of diffusion inwards of the ATP on the one hand, and the rate of hydrolysis by the actomyosin on the other. In the steady state, the concentration of ATP on the outside (denoted by C) is related to the concentration at the centre (denoted by I) according to the Meyerhof-Schulz formula:

$$C = Ar^2/4D + I$$

where A = rate of splitting, r = radius of the fibre, D = diffusion constant of ATP within the fibre.

Calculation shows that with a physiological concentration of ATP (about $5 \times 10^{-3}M$), the diameter of a fibre must not exceed about 6 μ if the concentration is to be the same in the centre. With a fibre bundle 500 μ in diameter, the concentration at the centre would be zero. In such a case, the fibers of a large central core remain stiff and play no part in the contraction, indeed even hinder it. Tension measurements have been successfully carried out with thin single fibers, for example by Briggs and Portzehl; here the fibers were 50 to 60 μ in diameter.

They showed a much higher tension production per square centimeter, about twice as great as with fibre bundles, presumably because the inactive core is much smaller. Even so, it was calculated that the ATP-free core would be 20 to 50% of the cross section; but it is much less likely than in a thicker fibre to be in a condition of rigor since it has available ample supply of ADP, a good plasticizer in the presence of enough Mg^{++}. It will be obvious that correlations of tension production and ATPase activity will present difficulties when fibre bundles are used, since the exact conditions of diffusion will vary from one bundle to another.

When on the other hand, comminuted glycerinated fibers or isolated fresh myofibrils some 2 to 3 μ in diameter are used, there is no diffusion problem, and enzymatic activity under different conditions can be accurately studied. Correlation of this ATPase activity with rate or degree of shortening is sometimes usefully made, but it is impossible to assess how far these mechanical effects can be taken as an indication of the power to produce tension.

ATPASE ACTIVITY OF MYOSIN AND ACTOMYOSIN

Purified myosin, free from actin, has ATPase activity, but actin has none. It is therefore assumed that the ATPase activity of actomyosin, though it shows some characteristic differences, is due to sites on the myosin component.

An important difference between actomyosin and myosin ATPase is that the former is activated by Mg^{++}, the latter not. This Mg^{++}-activation is seen with actomyosin at low ionic strength; above about 0.2, addition of ATP leads to dissociation so that then the enzymatic characteristics of myosin ATPase appear. Both ATPases are activated by Ca^{++}.

In view of the various indications that Ca^{++} and Mg^{++} ions are concerned with different sites on the myosin molecule, it is of interest to notice that the activation energy of the Mg^{++}-activated ATPase is much greater than that of the Ca^{++}-activated. Bendall (personal communication) has found an increase in rate of some 400-fold with the former on passing from 0°C. to 35°C.; with the latter, the increase was only some 20-fold. Hasselbach and Perry and Chappell have also noticed the high temperature coefficient of Mg^{++}-activated actomyosin. The ATPase activity of these two proteins has some special features which maybe of importance in contraction and relaxation.

THE HIGH INITIAL RATE OF A TPASE ACTIVITY

Weber and Hasselbach, using glycerinated fibers 30 μ thick, showed that the ATPase activity was at least twice as great during the first 15 sec. of reaction at room temperature as later—after about 100 sec., when a constant rate was reached. Accumulation of end products, decreasing concentration of ATP, impurity in the ATP, and contraction of the originally relaxed fibers were all considered ruled out as causes of the decrease in rate. The effect could be repeated if the hydrolysis was interrupted by Salyrgan inhibition, then reactivated by cysteine; it was in fact obtained only with fibers newly beginning to cause splitting.

The authors suggest that there may be slow reversible formation of an inactive enzymesubstrate complex of the sort described by Chance with catalase. The experiments described were done with Mg^{++} activation, though a case is given where Ca^{++}-activated myosin gave a similar but smaller effect. Bendall (personal communication) has found the effect to be consistently much smaller with Ca^{++} than with Mg^{++}-activation of myofibrils.

PHOSPHORYLATION OF THE ENZYME

On analogy with its other important biochemical reactions, it has often been suggested that, in ATP breakdown, the first stage is transfer of its terminal phosphate to some site on the myosin—or perhaps on the actin. If this mechanism does indeed operate, one should be able to find exchange between ADP^{32} and ATP in presence of the enzyme. Koshland*et al.* were unable to find such a reaction catalyzed by Ca^{++}-activated myosin or actomyosin. Thus if a phosphorylated protein intermediate is formed under these conditions, its existence must be transitory. This, as H. H. Weber has emphasized, is only to be expected, since a virtually irreversible, energy-yielding reaction would probably follow at once; moreover, any labeled ATP formed by the back reaction would be in a more favorable position for dephosphorylation than incoming ATP, since the former would be already in the neighborhood of the active sites.

Ulbrecht and Ulbrecht, using highly purified natural actomyosin or isolated myofibrils washed very thoroughly, showed that an exchange could readily be detected provided that Mg^{++} was the activator. However, this exchange was found to be independent of the presence of myosin—it continued in myofibril preparations from which the myosin had been removed. It is difficult to conclude that this means actin phosphorylation, since purified actin showed no exchange, although of course it can combine with myosin to give an actomyosin of normal ATPase activity and normal contractility in the gel form. The situation thus remains obscure.

SUBSTRATE INHIBITION

The existence under certain conditions of an optimal concentration of ATP for the ATPase activity of actomyosin has been realized since the work of Weber and Weber. They used glycerinated fibers and the phenomenon was further studied by Hasselbach and Weber. Recently, Perry and Grey have turned attention to this effect of overoptimal ATP concentration in connection with the mechanism of relaxation.

Using washed fresh myofibrils and working in the range of 2.5 to $10 \times 10^{-3} M$ for both ATP and Mg^{++}, they found that the ATP became inhibitory when its molar concentration exceeded that of the Mg^{++}. This relationship depended to some extent on the previous history of the myofibrils; it is well seen with myofibrils freshly isolated in 0.1 *M* KCl, and tested in a medium of about physiological ionic strength, 0.16. It does not, however, hold at lower concentrations of the reactants. Thus Geske*et al.* have found that with Mg^{++} concentrations below $5 \times 10^{-4} M$, the overoptimal ATP concentration shifts (with increasing Mg^{++}) first to *lower* ATP concentrations.

Perry and Grey found that marked substrate inhibition is characteristic of Mg^{++}-activated ATPase; it is very slight with Ca^{++} as activator. Very low Ca^{++} concentrations can abolish the substrate inhibition of the Mg^{++} activated system.

CONTRACTION AND ATPASE ACTIVITY

Weber and his collaborators have brought forward much evidence showing the close correlation between tension production in glycerinated fibers and their ability to dephosphorylate ATP. Thus Weber and Weber showed the effect of changing ATP concentration upon both tension and ATPase activity; the changes ran parallel both at room temperature and at 0°C.

Ulbrecht and Ulbrecht, with glycerinated fibers from smooth adductor muscles of *Anodonta*, found that the mechanical effects of ATP application and the ATPase activity showed the same temperature dependence. Bendall, working on the relaxing factor, found that as his preparations decreased the rate of shortening with glycerinated fibre fundles, so also they caused lowered ATPase activity. In all these cases, Mg^{++} was used as the ATPase activator and was supplied in the medium of the contracting fibers. Bowen and his collaborators, on the other hand, have described a number of cases which they consider can better be explained by correlating combination of ATP and actomyosin (rather than ATPase activity) with contraction. Thus they found conditions in which addition of Mg^{++} ions accelerated shortening of actomyosin filaments in presence of ATP, while homogenized preparations from the same material showed unchanged or decreased ATPase activity. On the other hand, Ca^{++} ions, which increased the enzymatic activity by a factor of 4, decreased the rate of shortening.

At this point, it is important to remember that the phenomenon of contraction is complex: ATP splitting may be an essential feature, but those who take this point of view would not regard it as the only essential requirement. Thus there seems good reason to believe that Mg^{++} ions play an important role (which cannot be filled by Ca^{++} ions) both in viscosity and in gelation changes; that they should play a specific part in the shortening process is therefore not unexpected. Indeed, Bendall (personal communication) has found, with well-washed glycerinated fibre bundles (about 200 μ in diameter) shortening under load, that there was virtually no work performance in 6 m*M* ATP when Ca^{++} was added up to 8 m*M*, although this Ca^{++} concentration can stimulate about maximal ATPase activity. Addition of 0.4 m*M* Mg^{++} to the bath (with or without Ca^{++}) led to good work production and ATPase activity. It seems, then, that the correlation is between Mg^{++}-activated ATPase activity and contraction.

Again, increasing the KCl concentration in the presence of Mg^{++} ions leads to accelerated shortening, while the enzymatic activity is depressed. Here, as Perry has pointed out, the increased ionic strength may, by reducing hydration of the fibers, improve their mechanical properties, quite independently of any effect on enzyme activity. Further, it has been found that the final degree of shortening of glycerinated fibre bundles varied with the concentration of ATP supplied, reaching in each case a constant value. The conclusion was drawn that the degree of shortening did not depend on the amount of ATP split, but on the concentration present, i.e. on the amount bound to the protein. But in these circumstances, the limiting factor in the hydrolysis rate in the interior of the fibers would be the rate of diffusion, itself dependent on the concentration.

There is nothing in these data to show that the shortening did not depend on the splitting rate; this is admitted but it is maintained that the comparison of the effects of Mg^{++} ions with those of Ca^{++} ions rules out this possibility. However, we have already shown the reasons for regarding this last argument as unsatisfactory. Many reagents reacting with SH groups are known to inhibit the ATPase activity of myosin. Their effect on contraction of actomyosin threads or fibers has always been found to be similarly or even more strongly inhibitory. In some cases, the ATPase activity of fibers after partial poisoning may be higher at room temperature than that of the unpoisoned fibre at 0°C.; yet the latter will contract, the former not. It would seem that, in the former, the energy released is not available for contraction.

It may be that SH groups are essential for reaction along the protein chain as well as for ATPase activity. To sum up, we may say that the evidence strongly favors the necessity of participation of Mg^{++}-activated ATPase in contraction; but the contraction process is complicated, involving more than this enzymatic activity.

MECHANISM OF THE ACTOMYOSIN ADENOSINE TRIPHOSPHATE INTERACTION IN CONTRACTION

We must now consider the stages of this interaction in more detail in relation to the mechanism of contraction. Ever since the realization of the part played by ATP, opinion has been divided as to whether its phosphate was liberated during the contraction or the relaxation phase; we shall deal with arguments put forward on both sides. It must be remembered that, until a few years ago, the view was generally held that contraction was produced by the folding of protein chains which ran throughout the muscle length. Recently, much evidence has accumulated that striated muscle contains two types of filament and that the individual filaments do not change in length as the muscle shortens. From these observations, the idea follows that the two types of filament slide over one another to produce the shortening. It has been suggested that a cyclic mechanism underlies this relative movement, linkages between the two types of filament (actin and myosin) being made and broken during contraction as the filaments pass each other.

THEORIES INVOLVING THE LIBERATION OF FREE ENERGY

Szent-Györgyi has for many years been a strong protagonist of this point of view, though it would seem that he has recently joined those who regard the contraction phase as the one needing direct provision of energy by ATP breakdown. He had supposed that upon stimulation, actin plus myosin forms actomyosin, and that ATP then combines with the myosin. In doing so, it enables a new endergonic link to be formed at the expense of the energy-rich phosphate bond. The establishment of this link has made certain joints pliable and these now fold, as a result of electrostatic attractions and repulsions along the chain. In order that relaxation may take place, the new energy-rich link must be broken, and this is done by the removal of ATP from the protein as ADP + H_3PO_4.

Avariation of this view is that of Riseman and Kirkwood; they postulated the phosphorylation of OH groups along the protein chain, which was then held extended by repulsion of the negative charges. On stimulation, dephosphorylation-took place, the charges were thus abolished, and the chain was free to contract. In relaxation, rephosphorylation of OH groups by ATP was necessary. On this theory, inorganic phosphate would actually be released during contraction, but the loss of free energy by the breakdown of ATP and formation of the phosphate ester linkage would occur during relaxation.

Next we come to the views of Morales and his collaborators, put forward during the last few years and summarized by Morales*et al*. These studies have been made on lightscattering by solutions of myosin B, i.e. the protein obtained by 5 to 24 hours' extraction, and considered by most workers to consist of a mixture of actomyosin and myosin, the former in much greater amount. Morales and his co-workers find the evidence for the presence of actin inconclusive, and consider the lightscattering phenomena to be due to myosin alone. The lightscattering method, as we have seen, is a very sensitive and rapid one for following changes in shape and size of protein particles. The effect of added ATP is a fall in the light scattering which persists for a time (during dephosphorylation of the ATP) and then there is a return to the original value. Below a certain ATP concentration, the degree of change depends on the concentration of the ATP added.

Further when the experiment was carried out in presence of 0.001 *M* Mg^{++} (which inhibits dephosphorylation at this high ionic strength), the lightscattering change (like the corresponding viscosity change) was unaffected in degree, but now was not reversed. This was interpreted to mean that the deformation of the particles depended on their combination with ATP and went on irrespective of ATP breakdown.

$$\mathrm{E} + \mathrm{ATP} \rightarrow E\,\mathrm{ATP} \rightarrow E + \mathrm{ADP} + \mathrm{P}$$

where *E* represents the enzyme after deformation. By studying the reaction at different temperatures, Morales and his collaborators consider that they have evidence that the light-scattering changes can go on without a change in molecular weight of the particles—i.e. without dissociation. In this, although using the same Zimm method, they are at variance with Gergely. In later work evidence for some dissociation was found by the Morales group, but the authors consider that only different states of aggregation of myosin itself were concerned when this happened.

Gergely and Martonosi on the other hand, have obtained further evidence in support of their interpretation, that dissociation of actomyosin is concerned, by preparing typical actin from 3 to 6 times precipitated myosin B. Morales and his collaborators concluded that the combination between protein and ATP is an energy-providing reaction (ÄF° = —6600 cal. per mole). They consider this deformation of the protein particles to be responsible for contraction and to depend upon the charge. They picture the degree of deformation as governed by a conflict between extensile electrostatic forces and contractile entropic forces—the particles gain configurational entropy on shortening, whether the shortening is by random coiling or by folding into an ordered form by reaction between certain groups. For relaxation, an energy-providing reaction is needed, to overcome the electrostatic effects of the bound ATP, e.g. the removal of the ATP from the protein by enzymatic breakdown to ADP and inorganic phosphate. The experiments that we have already described in which ATPase activity and contraction could be to some extent affected independently are cited as evidence for this point of view.

There is indeed abundant evidence that combination between ATP and myofibrillar protein in solution takes place and that "deformation" of the protein occurs before ATP dephosphorylation. Other workers have obtained results by the lightscattering method similar in many respects to those just described. The experiments employing the viscosimetric method cited earlier in Section II, B indicate the same thing. However, there seems to be, as we found in considering the effect of ATP on actomyosin solutions, good evidence that the "deformation" is a dissociation of myosin from actin.

When the effects upon a sol were compared with those on various forms of gel, we regarded the initial increase in extensibility as a comparable dissociation, a preliminary loosening of linkages putting the rigid gel into a fit state for contraction. In the resting muscle, this step is unlikely to be necessary, since electron micrographs seem to show the actin and myosin as distinct and separate filaments. But recent work has made it likely that any contraction involves a cyclic process of linkage-making and linkage-breaking, so that this dissocia-tion stage would be involved during the contraction phase (to make possible each further shortening step) and during relaxation.

FREE ENERGY OF ATP HYDROLYSIS

We have already discussed a good deal of experimental work which, taken at its face value, would seem to support this point of view. However, it is obvious that no enzymatic action by the actomyosin upon the ATP can take place without previous combination of the two, so that there is a logical flaw in any argument from a correlation between tension pro- duction or shortening and ATPase activity. The most striking evidence in support of this standpoint—that the energy of ATP hydrolysis is needed in the contraction phase—is that brought forward by H. H. Weber and by Bozler. We have already seen that a glycerinated fibre, made to contract in ATP solution, and remaining rigid and contracted when excess ATP is washed away, can be made to show a momentary relaxation by the addition of more ATP.

Weber and his collaborators now showed that this relaxation can be brought about in circumstances al- lowing no simultaneous energy provision. Thus ATP can bring about complete and lasting relaxation in the presence of the mercurial Salyrgan, which prevents its hydrolysis by poisoning the actomyosin ATPase. Salyrgan alone has no effect on the fibre. Inorganic pyrophosphate (0.0 15 M) also gave the same effect, although it is not a substrate for actomyosin ATPase. Bozler also got relaxation of ATP-contracted glycerinated fibers when 0.02 *M* pyrophosphate was added, and concluded that the relaxation process involves no large energy change. Bendall has got similar results with much lower pyrophosphate concentration (0.004 M), provided Mg^{++} (0.004 *M*) was also added.

This view of the timing of energy provision accords with A. V. Hill's finding that the relaxation phase (provided the muscle relaxes unloaded) is free from heat production. H. H. Weber has put forward a scheme which enables one to visualize possible mechanisms whereby the free energy of hydrolysis could be directly used in the contraction of the protein molecule.

Schemes such as this must inevitably be open to criticism as regards their details. Thus relaxation is pictured as brought about by hydrolysis of the ester linkage, and it is difficult to see how the relaxing effect of ATP fits in here. Morales and Botts have pointed out that the forces leading to establishment of covalent linkages are of very short range; further, such forces would become larger as the muscle shortened, while in fact the isometric tension developed becomes less as the muscle shortens. These latter objections are to some extent overcome when some such mechanism is envisaged (Weber, 1958) acting as part of a cyclic process. The evidence concerning the possibility of phosphorylation has already been considered.

THE SLIDING HYPOTHESIS

Electron micrographs of striated muscle have shown that two sets of filaments are present in regular hexagonal array and parallel to the long axis. Of these, the larger (diameter about 100 A.) have been identified as consisting mainly of myosin, the more slender (40 A. in diameter) as mainly of actin. The myosin filaments occupy the A band; the actin filaments run from the Z lines at the centre of the I band and into the A band as far as the beginning of the H zone. Upon contraction (to 6570% of the resting length), the A band remains of constant length, but the H zone and the I band shorten. These changes are best explained by a sliding of the actin filaments past the myosin into the A band, without folding of either type of filament unless shortening of the muscle is very great.

The evidence from low-angle diffraction diagrams obtained with living muscle supports the conclusion that the filaments themselves do not change in length. The problem then is to discover the mechanism whereby the actin filaments are drawn past the myosin filaments. In most of the suggested hypotheses, the contraction proceeds by stages. Sites on the actin filament can be pictured as oscillating by means of thermal agitation backwards and forwards past sites on the myosin filament with which it overlaps.

As a result of the stimulus, linkages are formed between the two sites and a certain degree of contraction occurs. When these links are broken by the arrival of fresh ATP, the actin sites are now within range of further myosin sites and the process is repeated further and further along the myosin chain. A schematic formulation based on what we know from studies *in vitro* of the interactions of actin, myosin and ATP.

1 (During the period of active contraction) M.ATP → P + ADP

2 (Formation of links, shortening stages) P + A → AM + P + free energy used as tension or work

3 (Loosening of links, finally relaxation) AM + ATP → M. ATP + A

Reaction: It is supposed that the effect of the stimulus is to cause a transphorylation which provides one of the proteins (myosin is suggested here but it might be actin) in a state ready to combine.

Reaction: The free energy of the energy-rich bond is used in some reaction or series of reactions leading to the formation of contracted actomyosin. This series of reactions might be, for instance, along the oblique connecting side chains suggested in one hypothesis or occupy the place of the "spring" in the hypothesis of A. F. Huxley. Formation of covalent bonds might be concerned as in Weber's formulation; electrostatic forces or hydrogen bonds might be concerned—we do not know.

Reaction: The loosening of the linkages would be due to the arrival at the sites of fresh ATP, by diffusion. On this formulation, ATP is concerned during contraction with both making and breaking of links; during relaxation, only with breaking of links. There is no requirement for energy provision during relaxation. The activation heat of Hill could arise:

- Possibly in connection with unknown reactions between stimulation and reaction (1);
- Possibly from some ATP dephosphorylation in the neighborhood of the sites, necessary to reduce it below a critical concentration;
- Perhaps from some heat wastage in reaction (1);
- From heat production connected with reaction (2), before enough links are formed to permit a mechanical response.

The shortening heat, only produced when the muscle actually shortens, would be derived from reaction (2), possibly also from reaction (3), though the latter contribution would have to be small as no heat is detectable during relaxation without load.

A very striking aspect of muscle metabolism is the constancy of the heat of shortening for a given muscle. Whether the muscle contracts slowly under a load or rapidly unloaded, the extra heat associated with a given amount of shortening is the same. The energy necessary for the accomplishment of the work is specially mobilized in proportion to the load and the shortening heat is not diminished to provide any part of it. Now as the muscle contracts more slowly when it is loaded, in these conditions more time is available and the same links might be formed and broken several extra times. We may surmise that this is the source of the extra energy which appears as work. Bendall has recently drawn attention to an assumption which is necessary if the sliding hypothesis is to be reconciled, as it must be, with the proportionality between shortening heat and distance shortened. It seems that as the actin and myosin filaments slide over each other, the number of reacting groups per unit of shortening must remain the same. Two alternative possibilities present themselves:

- On the actin filaments only the tips are reactive and groups situated here react with successive groups on the myosin;
- On the myosin filaments only the ends near the I band bear active groups and these react with site after site on the actin filaments. In either case a biochemical problem is presented, since so far as is known each type of filament is homogeneous along its length. Bendall makes the tentative suggestion that tropomyosin, quantitatively a minor constituent of the myofibril may have the rôle of protecting some parts of the major proteins from interaction.

The effort to go deeply into the sliding mechanism is only just beginning. A. F. Huxley has suggested a very interesting picture, involving oscillation of the myosin sites about an equilibrium position in the neighborhood of the actin sites, with certain postulations about the rate constants of the making and breaking of links, and the factors affecting these constants. Very reasonable quantitative agreement was obtained between the expectations worked out mathematically from this hypothesis and a great number of the thermal and mechanical observations on living muscle recorded in the literature.

ENERGY PROVISION IN THE LIVING MUSCLE

Using evidence from innumerable observations and experiments *in vitro*, we have built up the picture of an actomyosin-ATP machine depending for its energy supply on ATP dephosphorylation. We wish now to inquire how well this picture fits the requirements set by phenomena *in vivo* as regards the arrangements for the fuel supply. That is to say, how good is the evidence that ATP breakdown is *in vivo* the primary energyyielding reaction? It is not to be expected that, with a fresh, unfatigued muscle, such evidence will easily be produced, since, as we have seen, there are powerful enzyme systems in the muscle forrephosphorylation of ADP. Nevertheless, a number of lines of inquiry can be explored.

ATPASE ACTIVITY

In the first place, we may ask whether the activity of actomyosin ATPase under conditions *in vivo* is high enough to account for the energy production if all this is channeled through ATP breakdown. It seems that we must consider the behaviour of actomyosin ATPase, not that of free myosin ATPase. Muscle contains both Mg^{++} and Ca^{++}, but there is good evidence that only Mg^{++} activated actomyosin ATPase is concerned with contraction.

The results of Perry and Grey give a figure of about 0.3 μ, mole per milligram protein per minute for the Mg^{++}-activated ATPase of well-washed rabbit myofibrils at 20°C. The figures of Hasselbach are similar for finely-divided actomyosin gel at 23°C. Thus a figure of I to 2 μ moles per milligram protein per minute might be expected at 37°C., or 1 to 2 × 120 μ moles per gram muscle. This assessment takes no account of the high initial rate of inorganic phosphate liberation, which may be much more than double the rate estimated by observation over several minutes. If this high rate is substantiated as due to true ATPase activity, we must suppose it would operate in the conditions of discontinuous ATPase activity we have been contemplating. With 120 mg. of actomyosin per gram of muscle, the rate would then be at least 2 to 4 × 120 μ moles, or 2.4 to 4.8 × 10-4 moles per gram muscle per minute.

Bendall (personal communication) has indeed found initial values at 35°C. for rabbit myofibrils corresponding to 5-6 × 10-4 moles per gram muscle per minute. Mommaerts had calculated the utilization of energy-rich phosphate which would correspond to the maximum effort in human muscle; these calculations are based on the extra oxygen consumption, on the assumption that 5-6 energy-rich phosphate bonds are formed per molecule of oxygen used, and include corrections to allow for the actual amount of muscle tissue involved in the increased metabolism.

It seems that there may be a 500-fold rise in metabolism, and the metabolism of the energy-rich phosphate bond should reach 10 × 10-4 moles per minute per gram of muscle. The comparison of ATPase activity of rabbit muscle preparations with the muscle metabolism of man is open to criticism, since the small mammals show greater metabolic activity than the large. However, considering the roughness of all these calculations, the measure of agreement between the need and the supply is reassuring.

pH CHANGES IN LIVING MUSCLE

In experiments of Dubuisson, a glass electrode (timelag 2 to 3 sec.), in close contact with the muscle, was used to examine pH changes consequent upon a 4 sec. tetanus in the frog gastrocnemius. A series of three pH changes was seen-towards the acid side during shortening; then towards the alkaline side; finally another change to the acid side supervened.

The last two changes were mostly postcontraction. Dephosphorylation of ATP sets free acid, while dephosphorylation of creatine phosphate sets free base.The conclusion that the first change is due to ATP breakdown, the second to creatine phosphate breakdown, and the third to lactic acid formation, was supported by the effects of varying the initial pH of the muscle and of poisoning with iodoacetate.

CONSTITUENTS OF LIVING THINGS

The world, in itself, is a colourless, soundless, smell-less affair which none the less manages to titillate the senses so that they take on the garb of the more flashy secondary qualities. The intellectual tension caused by splitting and pulling apart the primary from the secondary qualities can be seen in the work of that honest underlabourer for science, John Locke. We can see the problem arising if we ask the question: What distinguishes atoms from the empty space they are supposed to occupy? Locke, closely following Robert Boyle, maintains that atoms possess primary qualities only. What are these qualities? Locke's list varies somewhat but we can take the following as typical: 'These I call *original or primary qualities* of body, which I think we may observe to produce simple ideas in us, viz. solidity, extension, figure, motion or rest, and number'.

Now if we look closely at this list then it seems evident that solidity is the only one of the primary qualities capable of doing the job of differentiating atoms from space. But what *is* solidity? He distinguishes it from hardness by saying 'that solidity consist in repletion, and so an utter exclusion of other bodies out of the space it possesses'; but, of course, to be replete is to be stuffed with, or full of, something or other and we still do not know what this something or other *is*.

The trouble now is (if we stay within Locke's scheme of primary and secondary qualities and within his empiricism) that the only kind of qualities with which we are directly acquainted and which are space-filling just are the secondary qualities. Colours and warmths, for example, occupy definite positions and take up shapes in space. In fact, Locke tries to save his empiricism by illicitly making the notion of solidity straddle the divide, on the one hand, between primary and secondary qualities and, on the other, between a directly perceivable quality and one that is not perceivable but which *explains* the space-filling and causal properties of bodies.

We can see this tension right at the beginning of his extended discussion of the 'simple idea' (that is, the simple sensation) of solidity: 'The idea of *solidity* we receive by our touch; and it arises from the resistance which we find in body to the entrance of any other body into the place it possesses till it has left it'. Already we can see him mixing up talk of the sensation of touch (what we might call a solidity-sensation) with what *causes* that sensation (resistance). (And, of course, if we are realists we should not accept that we get the notion of one body resisting another only from touch: we can, for example, just as well *see* bodies resisting each other). And at the end of the same section he makes the move of claiming that the purported explanatory notion of solidity is the very same as, or is very similar to, the simple *idea* of solidity which we get through touch: And though our senses take no notice of it but in masses of matter, of a bulk sufficient to cause a sensation in us, yet the mind, having once got this idea from such grosser sensible bodies, traces it further, and considers it, as well as figure, in the minutest particle of matter that can exist, and finds it inseparably inherent in body, wherever or however modified.

But the simple idea of solidity cannot be the same as the explanatory notion. The *idea* is just that, one of several, independent ideas or sensations whereas the *notion* of solidity is meant to explain several things about matter. And so he tells us: By this idea of solidity is the extension of body distinguished from the extension of space, the extension of body being nothing but the cohesion or continuity of solid, separable, movable parts; and the extension of space the continuity of unsolid, inseparable, and immovable parts. Upon the solidity of bodies also depend their mutual impulse, resistance and protrusions. This 'quality' of solidity obviously is not gained directly from experience and the pressing questions arise of what it possibly can be and how we can get to know what it is.

DISCOURSE AND REALITY LEVELS

Atomic theory, or its near relation, molecular biology, now claims to be able to give, in principle, a complete reductive account of sentient organisms. We are, in the first analysis, not much more than complex bundles of nucleic acid and protein and, in the last analysis, mere structured heaps of quarks and leptons. Many people have, naturally enough, disliked the idea that there is no place in the world for colours, scents and so forth and have liked even less the thought that nowhere, not even in their own heads, is there a mind to be found.

A move, popular among a number of biologists and philosophers who have reflected on these issues, has been to introduce the notion of *levels*. In this way they hope to retain realism about science while, at the same time, escaping reductionism and its implication. Thus, it might be said, although it is true that at one level, organisms are composed of quarks and leptons, it is also true that, at a different level, they are made of cells and, at a different level still, they may be conscious beings.

Very often such claims are combined with a notion of 'emergent evolution' in which it is held that quite new properties, such as consciousness, come into existence with, say, the increasing complexity of things. In any case, it is important to notice that these different levels—in which what gets squeezed out at a lower level may be restored at a higher—are taken as having independent, real existences. There is no question of the reduction, even in principle, of upper levels to lower.

It should at once be remarked that the ontological reductionist is not refuted by the fact that biologists, in their explanations, often mix descriptions and theories from different levels. It does not follow from the fact that the mixing of levels of *discourse* is frequent, and often necessary, that there must, therefore, be different *ontological* levels. James J. Gibson has tried to refute those reductionists who, 'impressed by the success of atomic physics, have concluded that the terrestrial world of surfaces, objects, places, and events is a fiction'. His answer to this is:

The world can be analyzed at many levels, from atomic through terrestrial to cosmic. There is physical structure on the scale of millimicrons at one extreme and on the scale of light years at another. But surely the appropriate scale for animals is the intermediate one of millimetres to kilometres, and it is appropriate because the world and the animal are then comparable. The reductionist's reply to this will surely be that mention of the supposed features of macroscopic objects, in explaining animal behaviour, is simply a methodological convenience—or even a methodological necessity.

For example, in the explanation of visual perception statements about the patterns of light and shade on objects may be mixed in with statements, at a different level, concerning the cells of the retina and the brain and yet others, at still another level, about photons. But one may take advantage of this methodological mix without being committed to the real existence of all the supposed levels. Similarly, there is more to the reductionist's case than justifies its dismissal, as Max Deutscher has recently undertaken, as a piece of obsessional thinking. Deutscher claims that the answer to the 'feeling' of reductive materialism 'lies in reminding ourselves of the plurality of facts and conceptual forms, and in regaining a free capacity to shift our perspectives'.

As barely stated this objection merely begs the question because, whatever may be the case concerning the plurality of conceptual forms or perspectives, the reductive materialist's argument just is that, in the sense required, there is *no* plurality of facts. As Stephen W. Hawking has put it: 'Since the structure of molecules and their reactions underlie all of chemistry and biology, quantum mechanics allows us in principle to predict nearly everything we see around us, within the limits set by the uncertainty principle.'

It would, says Deutscher, be as difficult 'to bring a volcano as to bring consciousness into a picture which allows no character or description except that contained within basic physics'. No doubt; but this incapacity, by itself, does not show there is anything more to a volcano than a collection of fundamental particles. It does not force the conclusion that: It is foolish to deny that volcanoes are composed of the elementary particles, and equally foolish to suppose that a story in terms of elementary particles presents the reality of things, whereas a story in the large scale about volcanoes and lava flows presents only a superficial appearance of the world.

Part of the problem here is that Deutscher does not spell out his 'story in the large scale'; and a reductive materialist would surely claim that the *full* story in the large scale, if only it were possible to tell it, would be in terms of the fundamental particles which together make up the volcano. As a matter of fact, scientists do try to give as fully a reductive account as possible of matter 'in the large' and in some cases, as with the quantum mechanical explanation and prediction of the properties of crystals and fluids, they have achieved remarkable success.

This last point deserves further emphasis because it also has force against Popper's claim that there have been very few successes for the proposed reductionist programme. The reductionist can, of course, here simply fall back on the fact of the great complexity of many things in the world. This is a hindrance to a full reductionist account being given of organisms, for example. But the reductionist can go further than this and point to the many successful achievements of the quantum mechanical theorists in recent years. These theorists have been able to explain and predict the properties of many materials, for example, because they have found ways of making simplifying assumptions concerning their constituent atoms.

These simplifying assumptions, themselves, depend on the presupposition that a full account of materials does depend on quantum-mechanical principles. Thus the wave functions of the all-important valence-electrons of an atom can be calculated to a good approximation by treating the nucleus and the inner electron shells as together constituting a 'pseudoatom' and this is an important step on the way to predicting important properties of crystals such as gallenium sulphide. As Cohen *et al.,* from whom I take this account, put it: Today the quantum mechanics of materials has become both conceptually and practically a simpler study than the study of the electronic structure of atoms having more than one possible valence-electron configuration. Through worldwide collaboration on pseudoatoms, alchemy, the original black art of materials science, is now well on its way to becoming one of the better developed parts of human knowledge.

Similar remarks could also be made about the theoretical and practical investigations of fluids. Unfortunately, the point is obscured by some authors who present us with a mishmash of levels of description and levels of actuality. Thus Steven Rose, partly to allay our fears about the implications of reductionism for the dignity of human beings, tells us 'that there are many levels at which one can describe the behaviour of the brain'; but he vacillates between talking about levels of description, which are in different 'universes of discourse', and discussing whether or not there can be causal relations 'between the point-set on one level and that on another'. The way he answers the latter question, by suggesting that we should be cautious in attributing causal rather than mere correlative relationships, suggests that he is now speaking of ontological levels so that, for example, the state of being in love occurs on one level and is correlated with molecular changes on another.

On the other hand, it seems that for Kathleen Wilkes there could be no question of a confusion of levels of language and ontological levels because ontological questions just *are* questions about language. In the midst of her spirited defence of physicalism about the mind-body problem she suddenly admits 'that the ontology, terms and explanations of common sense have nothing to do with physicalism'.

ONTOLOGICAL LEVELS

When we come to examine the notions of those thinkers who clearly espouse the doctrine of different levels there are further problems. There are variations in what are supposed to exist at different levels—qualities, laws, 'principles', things all have their advocates—and it may even be the case that, in a given hierarchical schema, different kinds of 'entities' occupy the various levels, e.g., things at one level and 'principles' at another. If we leave this problem aside there are still often difficulties in the interpretation of the concrete examples that various thinkers give. Thus consider the hierarchical list of levels proposed by Popper:

- Level of ecosystems
- Level of populations of metazoa and plants
- Level of metazoa and multicellular plants
- Level of tissues and organs (and of sponges?)
- Level of populations of unicellular organisms
- Level of cells and of unicellular organisms
- Level of organelle (and perhaps of viruses)
- Liquids and solids (crystals)
- Molecules
- Atoms
- Elementary particles
- Sub-elementary particles
- Unknown-sub-sub-elementary particles?

It is as if, with wedges and stars, the 'whole structure' of these things floats free from the fundamental constituents and, from its superior height, acts down upon them. I, for one, cannot make sense of this supposition.

We can see a related confusion in the work of those thinkers who wish to defend the notion of an infinite number of levels of physical being. David Bohm, for example, has urged that the history of atomic physics suggests that just as beneath the level of macroscopic objects there lies the level of atoms and molecules and beyond that again the level of 'fundamental particles', so, at least in principle, the physicist might go on and on discovering, without end, level after level. Now I think that in some sense the universe may be infinitely complex, and that, because of this, scientists may find eternal employment in unravelling the complexity; and it may also be that, despite their efforts, scientists will never get down to ontological bedrock; but it does not follow from this, nor indeed does the view make much sense, that there could exist an infinite number of ontological levels or strata. Indeed, I do not think there could exist more than one such level.

Once more the spatial metaphor of 'levels' in discussions of ontology is highly misleading. It is not the case, for example, that there is one set of entities, the protons, and *also* existing 'beneath' it, in some strange ontological space, a swarming sea of quarks which somehow supports its protonic flotsam and jetsam. The scientist, in finding out that protons are *really* triads of quarks in close interaction is discovering that there are no such separate things as protons—or, what comes to the same thing, that protons, in being made up of triads of mutually bound quarks, just *are* those threesomes in bondage.

To believe otherwise would, if the belief were rigorously followed through, involve awkward questions about the causal interactions between the protons and their 'constituent' quarks, remembering that the existence of the latter was, in the first place, postulated in order *fully* to explain the behaviour of those clumpings in space-time that were originally called 'protons'. Even if it were found that the quark theory did not provide a full explanation, that, for example, it were necessary to postulate some kind of 'membranes' to contain the separate triads of quarks, it would still be the case that the quarks and the 'membranes' all exist in causal interaction at the one level.

The attempt, then, to postulate an infinite series of levels of physical existence would be to believe in an infinite series of 'seemings to be' perched upon nothing at all.

POLANYI'S NOTION OF LEVELS

The reductionist's ability plausibly to deflect apparent counter-examples into the basket of merely methodological issues must strengthen his ontological claims. In fact, it is possible to show that in one recent influential positive defence of the idea of levels, that of Michael Polanyi, the argument depends for its apparent strength on an illicit mixing of ontological, and what may be loosely called methodological, claims. Because of this and other confusions the argument is difficult to state concisely. For convenience, I will first use Marjorie Grene's very clear formulation. Since they declare the processes of growth and heredity have been shown to be determined by a sequence of DNA molecules, biology has already been reduced, on principle, to bio-chemistry; the completion of the job is routine. Granted, however, that DNA has precisely the power they claim for it, its operation demonstrates, on the contrary, that biology is *not* reducible to biochemistry (and ultimately to physics).

What makes DNA do its work is not its chemistry but the order of bases along the DNA chain. It is this order which functions as a code to be read out by the developing organism. The laws of physics and chemistry hold, as reductivists rightly insist, universally; they are entirely unaffected by the particular linear sequence that characterizes the triplet code. Any order is possible physico-chemically; therefore physics and chemistry cannot specify *which* order will in fact succeed in functioning as a code.

This argument, which Grene thinks 'appears incontrovertible', would, if valid, beard the reductionist lion in his den. In fact, the argument very clearly is fallacious. If we look closely at the last sentence quoted above we perceive an equivocation. When Grene says, of the bits that make up the DNA molecule, that any order is possible physico-chemically she is thinking, as in the preceding sentence she states, of the *laws* of physics and chemistry; and scientific laws, by their nature merely allow *possibilities*—they do not, by themselves and in the absence of a statement of initial conditions, state what *must* happen.

When she concludes that 'physics and chemistry cannot specify *which* order will in fact succeed in functioning as a code' then, if her argument is to bear on the reductionist claim, she must now be thinking of the actual entities, including forces etc., that chemistry and physics describe. The conclusion must be that the structure of a particular DNA molecule is not due to the way its constituents as originally independent entities bumped up against, and reacted with, each other and this is borne out by the way Polanyi puts it: 'As the arrangement of a printed page is extraneous to the chemistry of the printed page, so is the base sequence in a DNA molecule extraneous to the chemical forces at work in the DNA molecule'. Because of this equivocation, the conclusion does not follow.

Once we do not feel forced to swallow the conclusion, because of the apparent reasonableness of the argument that led to it, we may feel free to examine it in its own right. When we do this we surely must see just how outrageous Polanyi's view is. The correct view would seem to be that it is just because the constituents of DNA are what they are, and because of the forces between them, that the laws of physics and chemistry 'are entirely unaffected by the particular linear sequence that characterizes the triplet code', i.e. that these laws 'allow' any *linear* sequence while ruling out other configurations.

The virtue of Grene's formulation of Polanyi's argument is that it allows us very quickly to see what is wrong with it. However, as Polanyi puts it, the argument comes embedded in a longer discussion, where there is a confusing mixture of ontological and 'methodological' issues, and itself then takes over this same confusion. It is precisely this that makes the argument seem so plausible. There is a hint of what is happening in Grene's account when she says, 'What makes DNA do its work is not its chemistry but the order of bases along the DNA chain. It is this order which functions as a code to be read out by the developing organism.' One's immediate response to this is that there is no distinction between the chemistry of the DNA molecule and the order of the bases any more than one distinguishes between the chemistry of acetic acid and the disposition of its component atoms.

One might say the structure of the molecule *is* part of its chemistry. The fatally confused step is when Grene goes on to say that the order of bases 'functions as a code' as if acting as a code involved something over and above acting chemically. In fact, her talk of the code being 'read out by the developing organism' should not be taken as a mere rhetorical flourish, as one might imagine it being used by a hard-line reductionist like Francis Crick, but is precisely where the whole confusion lies. This comes out in Polanyi's exposition where he talks at first of the DNA code which has 'information content' that, as he makes clear, is given by 'the numerical improbability of the arrangement'; but then immediately goes on to change his mind as to what 'information content' really is.

A printed page may be a mere jumble of words, and it has then no information content. So the improbability count gives the *possible*, rather than the *actual*, information content attributed to a DNA molecule; the sequence of the bases is deemed meaningful only because we assume with Watson and Crick that the arrangement generates the structure of the offspring by endowing it with its own information content. Polanyi has gone from speaking of 'information content' in a technical sense, as used in information theory, to a quite different use more akin to 'meaning' as when we speak of the meaning of a piece of prose. It is, of course, the first sense which is used in biological *science* and will be insisted upon by the reductionist as posing no threat to his reductionism.

This blatant mystery-mongering has its purpose. For Polanyi wishes to regard the DNA molecule as a mixture of blueprint and engineer which somehow *constructs* the living organism. 'Can the control of morphogenesis by DNA be likened to the designing and shaping of a machine by the engineer?' he asks, and answers in the affirmative. The point of this comparison is that he has earlier likened organisms to machines and has tried to show that even with ordinary machines, such as clocks, a reductive account of them cannot be given.

This is because a machine 'works under the control of two distinct principles. The higher one is the principle of the machine's design, and this harnesses the lower one, which consists in the physical-chemical processes on which the machine relies'. We might be tempted to regard the higher principle, that of the machine's design, which he describes as a 'boundary condition', as merely referring to the design of the machine as it is embodied in the machine, i.e. as the machine's structure; and then talk of the lower principles, i.e. the physical-chemical processes, as being 'harnessed' by the higher is at most a misleading metaphor. In one mood he does speak in this way as when he says, This harness is not unbreakable; the structure of the machine, and thus its working, can break down. But this will not affect the forces of inanimate nature on which the operation of the machine relied; it merely releases them from the restriction the machine imposed on them before it broke down.

If we are thinking, in this way, of the machine, in itself, then its structure does not harness its matter as a rider harnesses a horse—this sort of language can only be a florid way of saying that the structural arrangement of the material has to be given as an initial condition before, from it and the laws of mechanics etc., we can say what shall happen. There is no question here of higher and lower 'principles' and, therefore, no threat to the reductionist's position. This is disguised by the fact that Polanyi does not consistently refer to the machine and its structure, in itself, but also brings in the idea of the *constructor* of the machine. It is not now merely a question of the machine's structure harnessing the laws of nature—an innocuous idea if interpreted correctly—but the quite different concept of an *engineer* harnessing them.

The structure of machines and the working of their structure are thus shaped by man, even while their material and the forces that operate them obey the laws of inanimate nature. In constructing a machine and supplying it with power, we harness the laws of nature at work in its material and in its driving force and make them serve our purpose. Of course, in a sense, the engineer harnesses the laws of nature by bringing bits of matter into certain relations. But in saying this we are no longer thinking of higher and lower principles 'in' or 'of' the machine which is the conclusion Polanyi desires. If we are to speak, still, of two principles then it should be clear that there is now only one principle for the matter which makes up the machine and that the other resides with the engineer, which is only a complicated way of saying that there are two things in a certain relation to each other. Once seen in this light there seems no reason why a reductive account should not be given of the complex system: engineer plus machine.

In a loose way we can say that there is a confusion of ontological with methodological questions in that bringing in the engineer brings in a consideration of *interests*. The engineer, we might say, has a particular *interest* in the structure of a machine, either in bringing it about or in merely studying it. Similarly, a chemist may be interested in the atoms of which it is made. In fact, Polanyi, when speaking of a distinction he makes between machine type boundaries and test-tube type boundaries, seems to distinguish between the two according to whether our interest is in the boundary or the matter that is bounded; and that it is a purely methodological distinction is further suggested by his remark that 'By shifting our attention, we may sometimes change a boundary from one type to another'.

Also, in his *Personal Knowledge,* in what may plausibly be seen as a forerunner of the argument here criticised, Polanyi distinguishes between the physics and chemistry of machines and their 'operational principles', the latter being '*rules of rightness,* which account only for the successful working of machines but leave their failures entirely unexplained'. Here, again, I think we can easily detect questions of interest being confused with questions concerning the fundamental nature of things but I shall not pursue the issue further.

THE CONTRADICTION IN THE NOTION OF LEVELS

To assert the existence of different levels of 'principles' etc., unless this is taken as an obscure way of referring to different things interacting in a common environment, is to invite a criticism of the kind used to effect by the late Professor John Anderson and dubbed the 'two worlds argument' by John Passmore. This is that once two ontological levels, two distinct kinds of being, are distinguished then there is no way that they can, without contradiction, be brought together again in mutual interactions. The power of Anderson's argument is evident as it is employed against such metaphysical ideas as the ontological distinction between the platonic forms and the world of becoming or between God as the wholly other and the finite world he creates.

However, it can also be used where the proposed ontological 'split' is not, at least at first sight, as wide as this—including Polanyi's hierarchies if we take these to be ontological. For now it seems that in machines, for example, with the operation of higher order principles, the matter of which the machines are composed must act out of character. This is so because, in not taking these higher order principles as merely what occurs when bits of matter come into a certain relation with other things in the universe, e.g. engineers, they must be regarded as of the nature of those bits of matter.

By then saying that the lower order principles are 'open' to the higher, i.e. that they go on irrespective of the existence of the higher even although the latter also affect the action of the former, he is asserting that the matter of machines has both the character X and not-X. The argument has been put by P.H. Partridge for the general notion of emergent evolution. After arguing that the concept of causality as creative emergent evolution must take its sense from a supposed contrast with a preformationist idea of causality, where the effect somehow lies within the cause, he continues:

The theory of preformationism simply means that the effect is deducible from the cause alone, or is necessitated by the cause, it being the nature of a particular cause to produce such an effect. If, then, it is the nature of A to produce the one effect B, when A 'transcends itself by producing C, it is not acting according to its nature, but according to some other nature that is not its. Hence, the assumption that a thing both necessitates and creates implies that it is both itself and not itself, and this difficulty can be met not by a revision of logic, but, if we are unwilling to give up the notion of emergence as a special and peculiar occurrence, by the invention of some higher, mythological power asserting itself through the cause. If it is denied that the concept of emergent evolution involves the underlying assumption of these two different kinds of causality then the idea must be of one kind, which is creative in the sense that the effect is different from the cause, but with which one can nevertheless distinguish causes which have the most remarkable effects: effects which, as Samuel Alexander suggests, must be accepted with 'natural piety'.

There is here clearly a problem of how we distinguish between emergent and non-emergent effects although Alexander's own belief in the emergence of material things from naked space-time—as though space-time were a kind of unsuccessful porridge—may suggest there are some paradigms of the former. The point here seems to be that if such astonishing effects do burst forth in the cosmos then the rational course is not to accept them with 'natural piety' but to try to understand them and this involves finding out what their real causes are. An illustration of this point is afforded by a defence of the idea of emergence by the British Communist, John Lewis. Consider his account of the evolution of the DNA molecule:

What has to be noted is that:

- The process is wholly explicable in a succession of physical states operating under physical laws which
- Bring about the complete novelty of the DNA molecule possessing new properties and operating under its own laws.

Certainly the DNA molecule is unique but only in the sense that it is a unique spatial arrangement of its constituent atoms which, given what they are, will, in the relevant circumstances, fall out in the order they do. In this sense, the water molecule is also unique and, indeed, any molecule we care to think of. Of course, *in a sense,* the DNA molecule possesses new properties but only in the sense that any molecule possesses new properties, that is, those properties fitting for the kind of thing it is; but this sense does not exclude the fact that these new properties are fully explained in terms of the constituent atoms including the spatial etc. relations between them. In fact, the chemistry of one virus, X174, has been worked out completely, and the sequence of 5,375 nucleotides along its DNA, and what it 'codes' for, is fully known. The DNA of X174 did spring some surprises, connected with the details of the coding, but these were ordinary chemical surprises. It did not, for example, suddenly start whistling 'Waltzing Matilda'.

Suppose we do accept the idea of new qualities in evolution, does not this, despite Lewis's disclaimers, make their emergence entirely mysterious? Here the dualist might find his mark. Lewis claims that his scheme is not at all mysterious but this can only be in the sense that it is naturalistic and makes no appeal to 'supernatural' causes; but it certainly is mysterious in the sense that no *explanation* is possible of the emergent qualities nor, even, of why, for example, qualitatively identical DNA molecules should have the same emergent qualities. Creative the DNA molecules may be but they have all the spontaneity of the Guards on parade. The dualists might claim that their own account in terms of a creative, spiritual mind is superior because it can explain the emergence of new qualities. Or it may be that the dualist, like the mechanist, will reject the idea of emergence but, unlike the latter, accept the existence of undeniably mental events irreducible to mechanistic explanations.

It is the existence of such events which leads the dualist to dualism. We bake a cake and, as it cools, from nowhere a deep red icing settles on its surface. How do we explain this? By denying that it is a *magic* cake and instead saying that it is a *unique* cake where the flour and the currants have settled in just that one combination that allows for the *emergence* of red icing? Is this any more mysterious than the DNA molecule from nowhere obtaining brand new properties or the brain a coating of mentality? And if we try to lessen the mystery of the icing by postulating the existence of a ghostly chef, some departed Party member, perhaps, still anxious to do his bit, why would we think this more reasonable than conjuring up a god to paint on new qualities wherever in the universe they are needed?

If any content at all is to be given to ideas of 'emergence' or of 'hierarchies' then these ideas must be about what it is that exists at the one ontological level so that, for example, the dispute between the believer in emergent evolution and the mechanist is about whether or not vital forces or secondary qualities exist as well as quarks. This can be seen in Polanyi whose talk of mysterious emergent higher principles at one point becomes deflected into the advocacy of vitalism. Here the dispute is with orthodox Darwinians where Polanyi argues that rather than studying the origins of new populations one should try to account for the evolution of a single man by referring to that man's family tree which, he claims, 'includes everything that has contributed to the maturing of this human being'.

The point here is that, by denying the importance of commonly accepted environmental factors in human evolution, involving natural selection etc., Polanyi is, naturally enough, obliged to invent a different environment containing vital forces in order to account for this evolution; but these vital forces can then no longer be regarded as having a special ontological priority but must simply be bits of the universe interacting with other bits and, in particular, the complex chemical systems which are living organisms. This view, once teased out from the confusing talk of 'hierarchies' or 'emergence' and stated bluntly in its own right, can then be more coolly assessed.

TENSIONS

Modern biology, then, involves two fundamental metaphysical assumptions. The first is that, despite the useful retention of teleological forms of language, all causes of biological functional activity are efficient causes; and the second is that what, ultimately, exist to be studied by biology are structured collections of fundamental particles and their associated fields—*and nothing more*.

There is the further issue of whether, as is sometimes suggested, the category of causality should be given up altogether but, leaving that aside, these two general views are logically compatible. However, when we turn to the special Darwinian explanation of the origins of living things-which, of course, has played some part in the downfall of strictly teleological accounts—there is the problem of whether *it* is compatible with the strongly reductionist accounts of biophysics. The question is not often asked but needs to be taken seriously.

The theory of natural selection is, or at least, as originally conceived, was, concerned largely with explanation at the 'common-sense', macroscopic level and there is no guarantee that it will not suffer the fate of other theories of this kind that usefully miss the mark. As with the mechanics of gross bodies, the theory may be retained while the truth of the matter lies elsewhere, in among the strange interactions of what there ultimately is according to the quantum physicists.

Modern biology has had much to say about the fundamental constitutions of living things, including humans, and a good deal of it has been strongly reductionistic. If we leave aside the attempts of the logical empiricists to find logical relations of the appropriate sort between *propositions*—e.g. between the propositions of classical Mendelian genetics and those of modern biochemical genetics—then it seems that roughly two kinds of ontological reductionism have occupied thinkers in this area. On the one hand there are what we may generically refer to as naked ape types of theory and, on the other, there is the view of biochemists, such as Jacques Monod, that organisms (including humans) are nothing more than collections of molecules in certain complex relationships. Here the difference between the two kinds of theory appears as a difference in the level of explanation, taking this as relative to ordinary, everyday descriptions of the world.

Naked ape types of theory, including behaviouristic, socio-biological, etc. theories, remain close to everyday descriptions—it is just that one commonly noticed piece of behaviour is explained away as really being some other commonly noticed piece of behaviour as when it may be said that love is but lust with a fancy name. With the molecular biologists the world they describe seems quite different from the one that, in our non-speculative moments, we seem to inhabit. Colours, sounds, familiar objects all vanish.

If, for the time being, we accept this rough distinction between the two kinds of reductionist explanation, then we may notice that theories of the naked ape type—which try to pick on a few things to explain everything (both the few and the many things coming, in the first instance, from the level of the macroscopic)—are open to the sort of objection Robert Boyle used against the 'chymists' of his time. In the *Sceptical Chemist* Boyle argued against the then current theories of the 'elements' or 'principles' of matter (whether these be taken as earth, air, fire and water or, as in a rival scheme, salt, sulphur and mercury) that he could see no reason why these should be picked out of a myriad others, e.g. oils, as *the* elements.

Similarly, why should certain bits of behaviour, e.g. sexual behaviour, be picked out as *the* important bits when it is clear that there are many kinds of behaviour? By their very nature such theories are liable to leave important things unexplained. Boyle argued that the excessive narrowness of the chemical theories of his time, and their consequent lack of explanatory power, was heavily disguised by obfuscation of various sorts so that, for example, the chemists' elements took on all kinds of occult qualities.

We might expect, in our time, to detect similar techniques at work in behaviouristic theories, sociobiology, etc. where the existence of what does plainly exist, e.g. consciousness, altruism and so on, is either denied or in some way redefined so as not to constitute a threat to the theory. Often enough the redefinition of ordinary terms may serve to disguise various indecencies committed in the name of science, as when behaviourists use the term 'punishment' in a technical sense to designate actions which, in ordinary language, would properly be described as assault.

In particular, we should be on the lookout for the supposed basic elements of human nature taking on magical qualities so that, for example, the 'hunting instinct', supposedly exclusively possessed by men, becomes transmuted into the *naturally* male-dominated *pursuit* of knowledge; and where, with as much right to the carefree use of metaphor, feminists might claim that the *birth* and *nurture* of theories is an expression of the 'mothering instinct'. Such thinking, as many authors have pointed out, very often has a plainly ideological function.

In so far as a sharp distinction can be made between the different levels of theory such criticisms of naked ape type theories are just. However, the comparison with the theory of the elements suggests that a sharp distinction cannot be made between different levels of theory in this way if these are taken in relation to everyday descriptions.

What, on one view, might be seen as the attribution of merely occult qualities to the elements, in order to save appearances, may, on a different view, be more kindly regarded as an attempt to look behind the complex, changing world for ultimate principles of explanation. On this account, the chemists of Boyle's time are squarely in the western intellectual tradition of trying to find what it is that lies beyond the world of appearances and, in some sense, explains that world. It is merely that, unlike the atomists, say, they picked on the wrong kinds of things or qualities for their explanations. Similarly, hypotheses of hunting instincts etc. which, from one point of view, seem to be magically transmuted into activities usually seen as quite distinct from hunting etc. behaviour, may also be seen as attempts to penetrate behind appearances.

The latter interpretation can be illustrated by reference to pre-Socratic thought where the attempt was first made to find the basic stuff or stuffs of the world. Thales suggested it was water, perhaps, as Aristotle suggests in the *Metaphysics*, 'from observing that the nutriment of everything is moist, and that even heat is generated from moisture upon which it depends for its continued existence.

We might guess, from this suggestion of Aristotle, that the important transition in Thales's thought is from the more primitive question, 'Where do the things of the world come from?' which was hitherto answered by means of stories about the gods, to the question, 'What are the basic elements, or what is the basic stuff, of things?' If one asks the latter question then a satisfactory answer to it must involve choosing a candidate that, first, can be shown to be ubiquitous, to be in all things, and, second, has the right properties to explain the apparent diversity of things.

That water satisfied the first condition might have been suggested by what Aristotle gives as Thales's second reason for picking on it, 'that the seeds of everything have a moist character, and that water is the first principle in the character of all moist things'. Perhaps Thales thought the second condition was satisfied partly because he thought that one could observe, as a matter of fact, creatures emerging from watery substances such as the sea or semen; but partly, also, because he knew that water existed as various *phases* (liquid, solid and gas) and so, to that extent, could explain the apparent diversity of things in the world.

If this account of what Thales was about is anywhere near the truth then we can see that, like the seventeenth-century chemists or the twentieth-century naked ape theorists, he is picking on some one kind of thing to explain the characteristics of a multitude of things; and that, by doing this, he founds that intellectual adventure which endeavours to discover what lies behind the many perceived things.

To pick on water as the basic stuff of the world is, of course, to lay oneself open to the criticism, used by Thales's successors, that it simply will not explain the great diversity of objects. However, we can see with Thales two characteristic features of reductionist theories. The first is that the entities referred to by the reducing theory are picked as having certain general characteristics, e.g. the capacity of water to form the three phases of solid, liquid and gas, by which the entities referred to by the reduced theory or description are accounted for.

The second and correlative feature is that some characteristics only of the world that is said to be reduced are directly explained—thus Thales's theory may be said to explain, in a general way, why the world is made up of solids, liquids and gases but not the particular properties of, say, rocks, alcohol and mist. These two features stand in a potentially dialectical relationship. That, for example, there are features of the perceived world that the original hypothesis of water as the basic stuff does not explain might have led to modifications of the latter, so that water hypothesised as the basic stuff comes less and less to resemble water as we find it in rivers and seas; and, indeed, might have eventually been used to explain the properties of ordinary water along with those of other, everyday objects.

As the reducing theories gain in explanatory power, the entities they postulate become further and further detached from the common objects they explain. The other side of the dialectical coin is that, as these reducing theories develop in this way, they may come to affect our view of what it is about the perceived world that needs explaining.

Historically, because of its modest explanatory power, the water hypothesis was not modified in this way but was replaced by other equally mundane candidates. But as long as the basic principles or elements were taken as ordinary things of the world their appearance in explanatory theories was, like the crudities of the modern naked ape theorists, bound to be connected with viciously *ad hoc* arguments if these 'explanations' were even to appear to work.

What was required if this approach was to prove fruitful was the postulation of elements that, although bearing some analogy to the things of the world (and from where else could such an idea initially be got?), remained at some distance from it; and also that the properties attributed to these elements be such that they explain, in some profound way, the characteristics of the world as we perceive it. If such a theory is possible we need not expect it to be entirely correct straight off; but what we might anticipate is that the dialectical relation between it and the level of everyday observation, to which we earlier referred, will be fruitfully established. Such was the atomic theory of Democritus and Leucippus. This is not the place to go into the question of the original intellectual source of atomism or of its revival in the Renaissance. However, its explanatory power, compared with earlier theories of the fundamental constituents of the world, is evident. Atoms could be said to lie behind the perceived world in the apparently straightforward sense that they were too small to be seen; while, at least at first, they were held to have understandable properties, such as motion, shape, solidity, etc., which were taken from this same perceived world.

These supposed properties of atoms could explain many important features of the world, first in a qualitative way—as when Lucretius explains why water seeps through rock—and then in a precise mathematical way in explanation of such things as the gas laws. Once established the atomic theory sets going the dialectical process with the result that atoms lose many of their easily understood properties, such as hardness (so that no longer are both the hardness of rock and the malleability of clay, for example, explained in terms of lots of little hard things disposed in certain ways), and take on new and stranger ones. In this way atomic theory comes to explain more and more things in terms of properties which, although very strange indeed, are saved from the charges of vagueness and magic by their precise mathematical formulation and empirical testability.

CONSTITUENTS OF BLOOD

Blood is a tissue made up of cells in suspension and of plasma containing substances in solution. Forty-five percent of blood volume consists of the formed elements (red cells, white cells, and platelets). Fifty-five percent of blood volume is composed of plasma, which consists mainly of water, some proteins, and the various mineral salts necessary to life.

HOW IS BLOOD MANUFACTURED

Blood is assembled by the body from many different sources. With the exception of injections that may be necessary for treatment, all constituents of blood ultimately come from substances taken by mouth. The plasma is formed from water, proteins, and salts, all of which are absorbed from the digestive tract. The proteins are modified by the metabolism of the body into the type needed for the human species.The cells of the blood come from diverse sources. The various organs and tissues that collaborate in the manufacture of blood cells are known as the hemopoietic (blood-forming) system.

This includes the bone marrow, spleen, lymph glands, stomach, liver, and the reticuloendothelial system (a network of special tissue cells plus the lining of the blood vessels). The reticulo-endothelial cells are widely distributed throughout the body, but the most important sites are the lining of the bone marrow and spleen. They are phagocytic; that is, they can ingest micro-organisms and foreign particles of all sorts, thus removing them from the circulating blood.

FUNCTIONS OF THE BLOOD IN THE BODY

Blood, a complex fluid serving many purposes in the human body, preserves the internal environment required for normal activity of the living cells. Following are some of its main functions: Respiratory. Blood transports oxygen from the lungs to all the body tissues, and simultaneously carries carbon dioxide from these tissues to the lungs, where it is expired. (External respiration.)

Nutritive. Blood conveys food materials from the alimentary canal and food depots to all the tissues of the body. Excretory. Blood assists in the removal of waste products of metabolism by conveying them to the kidneys, where they are excreted in the urine. The maintenance of the water content of the tissues. Fluid lymph, containing oxygen and food materials, leaves the vascular channels and comes in contact with the tissues.Here it picks up carbon dioxide and other waste products and then flows back into the vascular channels. (Internal respiration.) The complex chemical processes of the body are enabled to take place because of the high dissolving and ionizing property of water in the blood and lymph.

Regulation of body temperature. The body owes its ability to regulate temperature largely to the water content of blood and of the tissue fluid. Heat from the deeper regions of the body is brought to the skin and lungs by the blood stream and dispelled.

Protective and regulatory. Blood and lymph contain complex chemical substances such as antibodies, antitoxins, and lysins—all of which protect the body against injurious substances. The circulating blood also brings hormones from the ductless glands into direct contact with the tissues that respond to such stimuli.

HOW MUCH BLOOD IS IN THE BODY

The average healthy individual has a blood volume equal to about 8 percent of his body weight. Thus the mean blood volume for a person weighing 110 pounds would be 4,400 ml.; for a person weighing 150 pounds, it would be 6,000 ml.; for a person weighing 200 pounds, it would be 7,600 ml. As 1,000 ml. is equal to 2.1 pints, it will be easy for one to estimate the number of pints of blood in his body. The volume for men will be slightly higher than that for women of the same weight because men have an agerage of 7½ percent more red cells than do women.

The amount of blood in the human body has been the subject of careful study. The methods used to calculate blood volume are based on this fact: When a known amount of concentrated substance is introduced into the blood stream, the extent to which the concentrate is diluted is a measure of blood volume. Substances used must be harmless in small quantities and easily identifiable in blood samples. Depending upon the type used, they attach themselves either to the red cell or to a plasma protein. Red cell volume has been measured by tagging the red cells with radioactive molecules; plasma volume has been measured by using a harmless blue dye. Information gained in this way can be used to calculate the whole blood volume.

The blood volume may vary within certain normal limits, depending upon the needs of the body. At rest, the circulating volume is at a minimum. But the blood volume may be increased by several conditions, including hot weather, high altitudes, muscular exercise, emotional excitement, and pregnancy. On the other hand, an abnormal decrease in blood volume may occur during anemia, hemorrhage, loss of plasma due to extensive burns, or by various forms of dehydration.

RED CELLS

Red cells, or erythrocytes, are small, solid, disc-shaped particles whose main function is to transport oxygen from the lungs to the tissues and to transport carbon dioxide from the tissues to the lungs. The red cells contain hemoglobin, a chemical substance that has an amazing affinity for oxygen and carbon dioxide. Compared to other cells in the body, red cells are incomplete and cannot reproduce themselves because they lose their nuclei shortly before being released from the bone marrow into the blood stream.

However, they are considered to be "alive" in the sense that they are able to perform their functions in the body for as long as 120 days. After that time they wear out and are then removed from circulation and destroyed by the spleen. The various components of protein and iron are salvaged for reuse by the body.

Red cells are created in vast numbers, mainly in the marrow of the short, flat bones of the body. It is estimated that they are worn out and removed from the circulation at the rate of 10 billion red cells an hour. The means by which the bone marrow must produce this same amount in the same period of time is part of the marvelous balance of supply and demand carried on by the normal physiology of the body. Red cells appear fragile in structure under the microscope but are actually very durable. They must be able to withstand a good deal of hammering about while being pumped through the blood vessels, and are able to squeeze through a capillary, which has a smaller diameter than the cell itself. In order to do this, the red cells must fold and bend in a flexible manner and be able to repeat the process many times.

While in the lungs the red cells are exposed to oxygen for only a brief time, but, owing to a chemical balance in the surrounding plasma, they become saturated with oxygen in 1 second. These oxygenated cells are then carried from the lungs to the heart and pumped out into the body. In the tissues, a delicate chemical change takes place in the plasma, and the oxygen is released for use. The carbon dioxide, created by the metabolism of the body cells, is picked up immediately by the red cell in exchange for oxygen and carried to the lungs. Here, another chemical change takes place in the plasma, a change enabling the red cell to release the carbon dioxide and almost instantly pick up more oxygen. This cycle is constantly repeated.

Red cells are very small, measuring about 7.5 microns in diameter and 1.95 microns in thickness. (A micron is about 1/25,000 of an inch.) It would require approximately 3,000 red cells lined up in their widest diameter to cover the space of an inch. Yet, because there are so many red cells in the human body, they would encircle the earth four times if lined up in a beadlike chain. In spite of the smallness of the individual red cell, the surface area of the combined cell mass is very large. In a man of average size, the surface area of the red cells represents a space as large as the surface of a football field.

Because one-fourth of the circulating blood is in the lungs at any one time, a combined cell surface area of 1,200 square yards is available to air. In 1 second, about 370 square yards of red cell tissue surface is exposed to the air, binding the oxygen to the hemoglobin. Only through these unique properties of the red cell, can manwith his relatively large size and weight-obtain enough oxygen to satisfy the needs of his complex system. From the heavy work of the muscles to the delicate functions of the brain, all cells and tissues depend on oxygen for life.

WHITE CELLS

The white cells, or leukocytes, are the protective cells of the blood stream. They are two or three times as large as the red cell and possess a nucleus. In a normal person, there are about 7,000 white cells per cubic millimeter * of blood (or about one white cell to 600 or 700 red cells). There are several varieties of white cells, and each variety has a different function. When a drop of blood is placed on a glass slide, spread into a thin film, and allowed to dry, it may be stained with a dye so that each type of white cell has a distinctive appearance and colour. White cells may thus be classified into two major groups: those that have easily visible granules in the cytoplasm (the body of the cell exclusive of the nucleus) and those that do not.

The most common type of white cell in the circulating blood has granules and is thus called a granulocyte. This variety comprises 60 to 70 percent of the white cells. The granulocvtes have the ability to migrate, like an amoeba, along the wall of a blood vessel and out of the capillaries into the body tissues, to fight bacteria. Also like an amoeba, they are able to engulf foreign substances. (This property is common to the granulocytes, monocytes, and reticulo-endothelial cells—all sometimes referred to as phagocytes.) When the body is threatened by infection, the granulocytes quickly gather at the point of invasion and set up a first line of defence. As an example, when you prick your finger and the bacteria enter and start to multiply, the body will mobilize many white cells at that point. They call sometimes be seen as a yellow accumulation called pus.

Lymphocytes, a second type of white cell, comprise 25 to 35 percent of the white cell population. These cells—formed by the spleen, lymph nodes, and other lymphoid tissue of the body—do not appear to be so aggressive in combating acute infection, but cluster in the tissues where chronic infection exists. There they form a barrier against the spread of infection to other parts of the body.

Lymphocytes may undergo certain chemical changes that are protective in nature, and they may have some function in the development of immunity. Monocytes, a third type of white cell found in the blood, are larger than the other white cells and relatively few in number. Like the granulocyte, they have the ability to migrate to the site of infection and ingest bacteria. Like a lymphocyte, they tend to wall off the infected area. During the healing process, they are capable of growing into another type of tissue repair cell, strengthening the weakened area.

The disease of leukemia is an abnormal, unrestrained overgrowth of white-cell-forming tissue. In other diseases, there may be an increase or decrease of white cells in the circulating blood. By obtaining a laboratory count of the number of white cells in a cubic millimeter of blood, a doctor has a valuable clue as to the state of health or disease of the body.

PLATELETS

Platelets are small, colorless, cell-like bodies that have an important function in the clotting of blood. They are a source of thromboplastin, a substance that is one of the links in the chain of clot formation. Platelets are liberated into the blood stream as fragments of a large parent cell in the bone marrow.

They are smaller than the red cell and fewer in number (one platelet to 10 to 20 red cells). The normal range in platelet counts is quite wide, varying from 200,000 to 400,000 per cubic millimeter of blood.

PLASMA

Plasma is the fluid part of the blood in which the red cells, white cells, and platelets are suspended. It is composed of 92 percent water and 7 percent proteins; the remaining 1 percent consists of fat, carbohydrates, and many mineral salts necessary for life. It also contains hormones, vitamins, and enzymes. Many of the plasma proteins have specific uses in the treatment of disease. The most well known of these plasma derivatives are gamma globulin, serum albumin, and fibrinogen.

Fresh frozen plasma is also prepared for the treatment of certain bleeding problems, especially hemophilia. Plasma is separated from whole blood by centrifugation; the red cells, being heavier, settle to the bottom of the blood-collecting container. The plasma can then be drawn off into separate containers and processed as fresh frozen plasma or sent to commercial firms for fractionation into specific derivatives.

BLOOD DERIVATIVES

Blood derivatives are those substances that can be separated from whole blood and used for specific purposes. Plasma itself is a component of whole blood. From the plasma there are a number of proteins that can be separated, and these fractions are usually called derivatives. Serum albumin, which in the body maintains plasma volume, is used for shock, low blood protein, and edema (the loss of water from the blood stream into the tissues). Antibodies against disease are concentrated in the gamma globulin.

Fibrinogen is one of the proteins essential for the formation of a blood clot. In addition to serum albumin, gamma globulin, and fibrinogen, a number of other plasma proteins are under study. The red cells left in the bottle after the withdrawal of plasma can also be used to great advantage in certain diseases.

BLOOD CLOT

The clotting of blood is the result of a delicate and complex series of biochemical events, some of which are not fully understood. This mechanism must satisfy two rigorous requirements if life is to be sustained:

- Blood must never clot while it is performing its functions within the intact circulatory system;
- Blood must always clot when any damage occurs that causes a break in the blood vessels. The body must be able to vary this complex response to bleeding ranging from a slight cut to a massive hemorrhage.

When tissues are damaged, thromboplastin is released from the nearby cells and platelets. This initiates a chain of reactions in which calcium plays an important part. The end result is that the soluble protein (fibrinogen) is converted to a gelatinous solid (fibrin). Fibrin forms in threadlike strands, making a webbed mesh in which the sticky platelets, red cells, and white cells are trapped, forming a jellylike clot. As the platelets distintegrate, they cause the fibrin web to shrink and have a constricting action on the walls of the torn blood vessel, causing it to clamp down and narrow the tear through which blood is escaping. This knowledge of clot formation is applied in the collection of blood for transfusions. The calcium, which plays a key role, is inactivated when it is mixed with a citrate solution in the blood container during collection.

ANTIBODY

An antibody is a substance in the plasma that reacts in a highly specific way with a foreign substance (called an antigen) when it is introduced into the body. Although this reaction is usually protective, it may sometimes result in injury to the body.

As an example of its protective function, if the body is invaded by a virulent strain of bacteria, two main lines of defence are set up. The white cells attempt to wall off and engulf the invaders; the bacteria, being a foreign substance, are antigens, and stimulate the reticulo-endothelial cells of the body to produce an antibody that will assist in repelling them. This antibody, contained in the globulin fraction of the plasma, may act on the bacteria in one of several ways. It may dissolve them; it may cause them to clump so that they cannot circulate freely through the blood stream; it may soften their protective membrane and inactivate them so that they will fall easy prey to the phagocytic white cells.

So it is that this type of antibody forms a protective agent against disease. Furthermore, as most of us are exposed to a large number of disease-causing microorganisms during our lifetime, we develop a concentration of various antibodies that may remain in the blood stream for a long time, even after we are well.

If we donate blood, the globulin fraction containing these protective substances can be separated from the rest of the plasma components and concentrated to about 25 times its normal strength. This concentrated substance, called gamma globulin or immune globulin, can then be injected into others to help protect them against certain diseases.

The antigen-antibody reaction also has an important application in giving blood transfusions. Red blood cells, if they are not the proper group and type, may act as antigens in the blood stream of the recipient. In this case, oddly enough, the antibodies in the A, B, AB, and O groups do not have to be developed by the introduction of red cells but exist in the blood stream in the natural state. In contrast to this, the antibodies to the Rh factor do not exist in the natural state and must be developed by a transfusion of incompatible Rh-type cells.

In the case of the ABO groups or the Rh types, the incompatible antigenic red cells are attacked by the circulating antibodies and caused to clump, or agglutinate, in the blood stream. If this reaction is strong enough, the red cells may be broken up, or hemolysed, by the antagonistic antibody. If enough of the transfused, incompatible red cells are clumped or broken up in the recipient's body, they have a damaging effect on the tubules of the kidney, and serious illness or even death may result from the inability of the kidney to excrete waste products. The steps by which these hazards are avoided will be discussed in a later section on laboratory work.

BLOOD GROUPS

Blood groups are inherited patterns into which human blood may be divided for scientific purposes. There are four major blood groups: A, B, AB, and O. * Groups A, B, and O were discovered by Dr. Karl Landsteiner in 1900, and a fourth, AB, was described by another scientist in 1902. Although the percentage of people belonging to each group tends to vary in different countries (as do other inherited characteristics, such as colour of eyes and hair), the following distribution represents the general spread in this country: 45 percent—group O; 40 percent—group A; 10 to 12 percent —group B; 3 to 5 percent—group AB.

About the turn of the century, there was great interest in learning how to use blood transfusions as part of the practice of medicine; but doctors were held back by a forbidding obstacle: while about half the transfusions given to treat illness had a remarkable therapeutic effect, the other half seemed to make the patient much worse and sometimes resulted in death. In seeking the clue to this riddle, Landsteiner performed a series of tests in the course of which he placed the red cells of one person in suspension in the serum of another person. From this crucial experiment, he noted that in a number of the combinations of red cells and sera, the red cells remained undamaged and freely suspended; in other combinations, the red cells were clumped together in sticky clusters—with resultant damage to the cells.

From this evidence, he deduced that certain combinations of cells and sera were incompatible and that it was these unfortunate combinations that were causing illness and death in some patients following transfusion (transfusion reaction).Further investigation of the factors causing clumping, or agglutination, of the red cells when exposed to incompatible sera revealed that there are two main antigenic substances (agglutinogens) in the red cells and two main antibodies (agglutinins) in the plasma. The antigens in the red cells were named A and B, and the antibodies in the plasma were given names to correspond with the group of red cells that they would agglutinate.

Therefore, the plasma antibody causing group A cells to clump was called anti-A; the plasma antibody causing group B cells to clump was called anti-B.

As noted in the discussion on antibodies, the anti-A and anti-B substances are contained in the globulin protein of the plasma. These basic facts enable us to understand the four simple combinations that are the bases for the major blood groups:

- *Group A blood:* Has A substance in the red cells, and anti-B substance in the plasma.
- *Group B blood:* Has B substance in the red cells, and anti-A substance in the plasma.
- *Group AB blood:* Has both A and B substance in the red cells, and neither anti-A nor anti-B substances in the plasma.
- *Group O blood:* Has neither A nor B substance in the red cells, and has both anti-A and anti-B substances in the plasma.

It will be noted that each blood group is named for the type of antigenic substance that is an inherited characteristic of the red cell.

DETERMINATION OF BLOOD GROUPS IN THE LABORATORY

Blood groups are determined in the laboratory by a simple technique utilizing the known facts about the A and B substances in the red cells, and the anti-A and anti-B substances in the plasma. On the left side of a clean glass slide, the technician places a drop of serum known to contain anti-A substance; on the right, a drop of serum known to contain anti-B substance. He then places a drop of blood (obtained from the person whose group he wishes to determine) into each of the specimens of anti-A and anti-B sera. After a minute or two, the technician can observe one of four possible events take place by which he can determine the blood group:

- If the anti-A serum alone causes the red cells to clump, the blood sample is labeled *group* A.
- If the anti-B serum alone causes the red cells to clump, the blood sample is labeled *group* B.
- If *both* anti-A and anti-B sera cause the red cells to clump, the blood sample is labeled *group* AB.
- If *neither* anti-A nor anti-B serum causes the red cells to clump, the blood sample is labeled *group* O.

The clumping, or absence of it, is clearly visible to the naked eye in a few moments. Thus, in a well-organized laboratory the technician can determine the groups of a large number of samples in a few hours' time.

For safety, the results of blood group determination are checked in the laboratory in another procedure called proof of blood group. This utilizes samples of known group A and group B red cells, against which samples of the unknown serum are tested. Once again, four observable results may be noted: the group A cells alone may be clumped, the group B cells alone may be clumped, both group A and B cells may be clumped, or no clumping at all will occur. One can easily work out the blood group that would occur with each of these events. Still another use is made of these facts when the bottle of blood, properly identified as to group, reaches the hospital and is ready to be transfused into the injured or ill patient. Even though the blood is the same group as the patient's, another precaution must be taken to ensure that bloods of donor and recipient are compatible.

This is a laboratory test, known as crossmatching. A drop of the donor's red cells is mixed with a drop of the recipient's serum; this is known as the major side of the crossmatching test. The mixture of recipient's cells with donor's serum is known as the minor side of the test. If agglutination occurs on the major side, the blood is incompatible and must never be given because the donor's cells would be agglutinated and hemolyzed in the recipient's blood stream, causing dangerous transfusion reaction.

If agglutination occurs on the minor side, as it will when group O blood is given to A, B, or AB recipients, it does not mean that the blood cannot be given. At the present time, however, it is the practice to give group-specific blood (group A to group A recipients, B to B recipients, and so on) wherever possible and to use group O donors as universal only in emergencies.

HOW ARE BLOOD GROUPS INHERITED

Blood groups are inherited as dominant characteristics according to the Mendelian law: for every inherited characteristic, including blood group, there is a pair of genes, one contributed by each parent.

The blood group that a child may inherit is therefore dependent on the combinations resulting from genes contributed by both parents. Owing to intermarriage of people with different blood groups, a variety of combinations can result; but if the parents' blood groups are known, those possible for the children can be predicted.

MEDICOLEGAL IMPLICATIONS OF BLOOD GROUPS

Blood groups are inherited and never change. For this reason, they may have legal importance in identifying bloodstains or determining parentage. But the evidence so provided is useful only in a negative sense. For example, if a murder suspect had his clothing stained with group A blood, and the victim was of group O, this is evidence that the suspect was not stained by the victim's blood.

However, even if the stains proved to be group O, it still would not be conclusive evidence, for the suspect may have had his clothing stained with the blood of any person who was group O. This logic also applies to paternity cases. The determination of paternity in disputed cases is very complicated. Such cases must be studied by experts in forensic medicine.

THE Rh FACTOR

The Rh factor is an inherited antigenic substance in the red cells (similar in nature to the group A and B substances) that occurs in 85 percent of the white population of this country. Those who possess this factor are said to be "type Rh positive." Those who do not possess it are "type Rh negative." It occurs in about the same percentage in all blood groups; that is, of 100 group A bloods, approximately 85 will be Rh positive and 15 Rh negative; of 100 group B bloods, approximately 85 will be Rh positive and 15 Rh negative. These same percentages hold for all four blood groups.

The Rh factor was named in 1940 by Landsteiner and Wiener, who were doing work with the red cells of the rhesus monkey. They discovered that the antiserum that agglutinated the cells of the rhesus species also agglutinated 85 percent of the donors they examined; it failed to agglutinate the other 15 percent. The word rhesus was shortened to Rh in discussions of this new factor and thus entered the literature. In 1939 Levine and Stetson reported a severe transfusion reaction in a woman delivered of a dead baby, and discovered that her blood would agglutinate 80 percent of donors who were of her own blood group. They assumed that the infant had inherited from the father a blood factor that was not present in the mother's blood. Small quantities of the fetal red blood cells had seeped through the placental membranes and caused antibody production in the mother. The brilliant scientific work by which these groups of research workers correlated their separate findings and laid the foundations for the important work on the Rh factor is worthy of additional reading.

SIGNIFICANCE OF THE Rh FACTOR

The Rh factor is of significance in blood transfusion and may be significant in childbirth. The Rh substance can cause a reaction if injected into the blood stream of a person who is Rh negative; that is, his body will have a tendency to break down the Rh-positive cells. This may clog the kidney with the residue of broken-down red cells and, if severe, can result in death.

However, the first tranfusion may not cause this reaction because the recipient may become only sensitized; in other words, he develops anti-bodies—but not enough to produce severe symptoms. However, having been sensitized, he might suffer a severe reaction from subsequent transfusions. In pregnancy the Rh-positive child of an Rh-positive father may be carried by an Rh-negative mother. In a small percentage of these cases, Rh-positive cells from the infant's blood stream cross the placental barrier membrane and seep into the mothers' circulation.

Because these cells of the infant carry an antigenic substance, Rh, they may evoke in the mother's body an antibody response against the cells. This has little effect on the mother because the amount of the infants' cells in her blood stream is very small compared with the volume of her circulating blood.

However, if the antibodies to the Rh-positive cells generated by the mother are carried back into the blood stream of the infant, they may do a great deal of damage. The concentration of the antibodies in the small body of the infant is so great that it may damage a large percentage of the red cells. Depending on the degree of severity with which this happens, the infant at birth may be merely jaundiced, a condition that may clear up in a few days; he may be quite ill, and to save his life may require an exchange transfusion (removal of baby's blood and simultaneous replacement with Rh-negative blood). In a very few cases, the baby is born dead.

Because the Rh factor has received a great deal of publicity during the past few years, many women fear the possibility of giving birth to an infant affected by hemolytic disease of the newborn (erythroblastosis). Actually, much of this apprehension is unfounded. It should be understood that Rh is a normal inherited characteristic, just as colour of eyes and hair. Again, marriages with an Rh-negative woman and an Rh-positive man may occur in approximately 13 percent of marriages (15% x 85% = 13%).

Most women who are Rh negative do not produce anti-Rh antibodies, and those who do usually have at least two babies that are unaffected. Even in marriages with an Rh-negative woman and an Rh-positive man, only 1 in about 26 has an infant affected, and many of these children only mildly so. With the present knowledge about blood, there is a much better outlook for detecting any Rh incompatability and treating the affected infant.

3

Biogeochemical Cycle

ECOSYSTEM

An ecosystem is a group of living and non-living components interacting together on a given physical landscape. The size of an ecosystem is arbitrary and could be as small as a few square centimeters if you are looking at a soil microbial ecosystem; as large as thousands of square kilometers if you are looking a biome like the Great Plains ecosystem; or a few hectares if you are looking at a single forest stand ecosystem.

FORESTED ECOSYSTEM

An ecosystem model is an accurate but simplified representation of an ecosystem that can be very useful in thinking about or simulating the actions of a real ecosystem. Because any ecosystem has many different but interrelated components, the best way to understand the system is to break it down into its component parts. To get an introduction to a very simplified forest model, which gives participants and introduction to how a hardwood forest ecosystem works before and after exotic earthworms invade.

- *Step One*: The first step in building a graphical model of a hardwood forest ecosystem is to identify its major components. The components of any ecosystem are those physical things that contain energy and nutrients.

  A forested ecosystem, by definition contains trees, so that is our first component. In addition to the various species and layers of trees in a forest, there are other distinct ecosystem components. The *understory* contains most of the visible plant life found between the sapling layer and forest floor. We will add those two components to our forest ecosystem model later
- *Step Two*: Once you have identified the components of your ecosystem model, you need to define the *processes* that connect the components. This is graphically done by using arrows to indicate the flow of nutrients or energy among the different ecosystem components. The

components of our ecosystem model are now connected by processes that result in the movement of energy or nutrients among the components.

One thing to notice in our ecosystem model is that there is no process connecting the *trees* component with the understory plants component. This is because there are no substantial processes that result in the flow of nutrients directly from a tree to an understory plant or visa-versa. There would be important relationships between the trees and understory plants. Trees provide shade to the understory plants. But remember, in an ecosystem model, only processes that result in flow of energy or nutrients are represented. Now let's add the animals and the people components to our ecosystem. energy and nutrients flow from the trees and understory plants to the animals when they eat the leaves, twigs and buds of trees or graze on understory plants; and when the animal excrete waste products or die, energy and nutrients are returned to the forest floor component.

- *Step Three*: Determine the major inputs and outputs of your ecosystem. As you are building your ecosystem model, one thing to think about is whether your ecosystems could be opened or closed. A closed ecosystem is one that has no inputs of energy or nutrients from outside the ecosystem and no outputs of energy or nutrients leaving the system. The earth is an example of a closed ecosystem with respect to nutrients and an open ecosystem with respect to energy All the nutrients that have ever been on earth are here and simply continue to cycle, there are no additions or losses. However, the earth is constantly getting inputs of energy from the sun and simultaneously radiating energy back.

  The earth ecosystem and has no inputs or outputs of nutrients which are constantly recycled within the global ecosystem, while the earth has both inputs and outputs of energy that are in a relatively stable equilibrium.

  Just as the Earth ecosystem is closed with respect to nutrients, unmanaged earthworm-free hardwood forest ecosystems are often very nearly closed nutrient ecosystems that there are very few inputs or outputs of nutrients. Rather the nutrients are constantly recycled among the various ecosystem components. In contrast, most agricultural ecosystems require nutrient inputs from outside to function properly some typical inputs and outputs of nutrients and energy for forested ecosystems include evapotranspiration, nutrient leaching, sunlight and rain.
- *Step Four*: Once you have identified the components, processes and major inputs and outputs in your ecosystem model, then you can

begin to add the actual values to these parts of your ecosystem by measuring them. The amount of litter that falls to the forest floor each year (a process), what the biomass of trees is in a given forest (a component), how much light reaches the forest over a growing season (an input), or how much nitrogen leaches from the forest (an output). In the case of leaf litter, you can put out trays in the forest and after all the leaves have fallen for the year, dry and weight the leaf liter in your trays. They you can use that value to calculate an estimate of the total leaf litter for your forest

- *Step Five*: Use your ecosystem model to think about how changes can cascade through an ecosystem or to ask specific questions that can be answered with further research. When the major components, processes and inputs and outputs of an ecosystem are understood, then you can use the model how changing one part of the ecosystem affects other parts. If you harvest trees from your forest, that will decrease the amount of leaf litter reaching the forest floor each year which may lead to decreases in available nutrients for understory plants.

An ecosystem that is in equilibrium doesn't gain or lose nutrients. Researchers may monitor soil nutrient levels for many years after trees have been harvested how the real forest behaves compared to what they thought might happen based on their forest model, their understanding of how the forest works.

If the results in the real forest are very different than those predicted by their model, then they know that they don't have full understanding of how their forest works and they may go back to try to improve their model. In general, when building a model, you start simple and add more detail as needed. To get an introduction to a very simplified forest model which gives participants and introduction to how a hardwood forest ecosystem works before and after exotic earthworms invade.

ENERGY FLOW THROUGH THE ECOSYSTEM

Both energy and inorganic nutrients flow through the ecosystem. We need to define some terminology first. Energy "flows" through the ecosystem in the form of carbon-carbon bonds. When respiration occurs, the carbon-carbon bonds are broken and the carbon is combined with oxygen to form carbon dioxide. This process releases the energy, which is either used by the organism or the energy may be lost as heat. The dark arrows represent the movement of this energy. The other component are the inorganic nutrients. They are inorganic because they do not contain carbon-carbon bonds. These inorganic nutrients include the phosphorous in your teeth, bones, and cellular membranes; the nitrogen in your amino acids; and the iron in your blood.

The movement of the inorganic nutrients is represented by the open arrows. Note that the autotrophs obtain these inorganic nutrients from the inorganic nutrient pool, which is usually the soil or water surrounding the plants or algae. These inorganic nutrients are passed from organism to organism as one organism is consumed by another. Ultimately, all organisms die and become detritus, food for the decomposers. At this stage, the last of the energy is extracted and the inorganic nutrients are returned to the soil or water to be taken up again. The inorganic nutrients are recycled, the energy is not.

Many of us, when we hear the word "nutrient" immediately think of calories and the carbon-carbon bonds that hold the caloric energy. When writing about energy flow and inorganic nutrient flow in an ecosystem, you must be clear as to what you are referring. Unmodified by "inorganic" or "organic", the word "nutrient" can leave your reader unsure of what you mean. This is one case in which the scientific meaning of a word is very dependent on its context. Another example would be the word "respiration", which to the layperson usually refers to "breathing", but which means "the extraction of energy from carbon-carbon bonds at the cellular level" to most scientists (except those scientists studying breathing, who use respiration in the lay sense).

To summarize: In the flow of energy and inorganic nutrients through the ecosystem, a few generalizations can be made:

- The ultimate source of energy (for most ecosystems) is the sun
- The ultimate fate of energy in ecosystems is for it to be lost as heat.
- Energy and nutrients are passed from organism to organism through the food chain as one organism eats another.
- Decomposers remove the last energy from the remains of organisms.
- Inorganic nutrients are cycled, energy is not.

FOOD CHAINS AND WEBS

A food chain is the path of food from a given final consumer back to a producer. For instance, a typical food chain in a field ecosystem might be:

Grass → Grasshopper → Mouse → Snake → Hawk

Note that even though I said the food chain is the path of food from a given final consumer back to a producer we typically list a food chain from producer on the left to final consumer on the right. Note to international readers: In Hebrew or Aramaic, or other languages which are read right-to-left, is it customary to list the food chains in the reverse order? By the way, you should be able to look at the food chain above and identify the autotrophs and heterotrophs, and classify each as a herbivore, carnivore, etc. You should also be able to determine that the hawk is a quaternary consumer. The real world, of course, is more complicated than a simple food chain.

While many organisms do specialize in their diets (anteaters come to mind as a specialist), other organisms do not. Hawks don't limit their diets to snakes,

snakes eat things other than mice, mice eat grass as well as grasshoppers, and so on. A more realistic depiction of who eats whom is called a food web. It is when we have a picture of a food web in front of us that the definition of food chain makes more sense. We can now a food web consists of interlocking food chains, and that the only way to untangle the chains is to trace *back* along a given food chain to its source.

The food webs are *grazing food chains* since at their base are producers which the herbivores then graze on. While grazing food chains are important, in nature they are outnumbered by *detritus-based food chains*. In detritus-based food chains, decomposers are at the base of the food chain, and sustain the carnivores which feed on them. In terms of the weight (or biomass) of animals in many ecosystems, more of their body mass can be traced back to detritus than to living producers.

Pyramids

The concept of biomass is important. It is a general principle that the further removed a trophic level is from its source (detritus or producer), the less biomass it will contain (biomass here would refer to the combined weight of all the organisms in the trophic level).

This reduction in biomass occurs for several reasons:

- Not everything in the lower levels gets eaten
- Not everything that is eaten is digested
- Energy is always being lost as heat

It is important to remember that the decrease in number is best detected in terms or biomass. Numbers of organisms are unreliable in this case because of the great variation in the biomass of *individual* organisms. For instance, squirrels feed on acorns. The oak trees in a forest will always outnumber the squirrels in terms of combined weight, but there may actually be more squirrels than oak trees. Remember that an individual oak tree is huge, weighing thousands of kilograms, while an individual squirrel weighs perhaps 1 kilogram at best.

There are few exceptions to the pyramid of biomass scheme. One occurs in aquatic systems where the algae may be both outnumbered and outweighed by the organisms that feed on the algae. The algae can support the greater biomass of the next trophic level only because they can reproduce as fast as they are eaten. In this way, they are never completely consumed.

It is interesting to note that this exception to the rule of the pyramid of biomass also is a partial exception to at least 2 of the 3 reasons for the pyramid of biomass. While not all the algae are consumed, a greater proportion of them are, and while not completely digestible, algae are far more nutritious overall than the average woody plant is (most organisms cannot digest wood and extract energy from it).

Biological Magnification

Biological magnification is the tendency of pollutants to become concentrated in successive trophic levels. Often, this is to the detriment of the organisms in which these materials concentrate, since the pollutants are often toxic. Biomagn-ification occurs when organisms at the bottom of the food chain concentrate the material above its concentration in the surrounding soil or water. Producers, as we saw earlier, take in inorganic nutrients from their surroundings. Since a lack of these nutrients can limit the growth of the producer, producers will go to great lengths to obtain the nutrients.

They will spend considerable energy to pump them into their bodies. They will even take up more than they need immediately and store it, since they can't be "sure" of when the nutrient will be available again (of course, plants don't think about such things, but, as it turns out, those plants, which, for whatever reason, tended to concentrate inorganic nutrients have done better over the years). The problem comes up when a pollutant, such as DDT or mercury, is present in the environment. Chemically, these pollutants resemble essential inorganic nutrients and are brought into the producer's body and stored "by mistake". This is the first step in biomagnification; the pollutant is at a higher concentration inside the producer than it is in the environment.

The second stage of biomagnification occurs when the producer is eaten. Remember from our discussion of a pyramid of biomass that relatively little energy is available from one trophic level to the next. This means that a consumer (of any level) has to consume a lot of biomass from the lower trophic level. If that biomass contains the pollutant, the pollutant will be taken up in large quantities by the consumer. Pollutants that biomagnify have another characteristic. Not only are they taken up by the producers, but they are absorbed and stored in the bodies of the consumers.

This often occurs with pollutants soluble in fat such as DDT or PCB's. These materials are digested from the producer and move into the fat of the consumer. If the consumer is caught and eaten, its fat is digested and the pollutant moves to the fat of the new consumer. In this way, the pollutant builds up in the fatty tissues of the consumers. Water-soluble pollutants usually cannot biomagnify in this way because they would dissolve in the bodily fluids of the consumer. Since every organism loses water to the environment, as the water is lost the pollutant would leave as well. Alas, fat simply does not leave the body. The "best" example of biomagnification comes from DDT. This long-lived pesticide (insecticide) has improved human health in many countries by killing insects such as mosquitoes that spread disease

On the other hand, DDT is effective in part because it does not break down in the environment. It is picked up by organisms in the environment and incorporated into fat. Even here, it does no real damage in many organisms (including humans). In others, however, DDT is deadly or may have more insidious, long-term effects. In birds, for instance, DDT interferes with the deposition of calcium in the shells of the bird's eggs.

The eggs laid are very soft and easily broken; birds so afflicted are rarely able to raise young and this causes a decline in their numbers. This was so apparent in the early 1960's that it led the scientist Rachel Carson to postulate a "silent spring" without the sound of bird calls. The banning of DDT, the search for pesticides that would not biomagnify, and the birth of the "modern" environmental movement in the 1960's. Birds such as the bald eagle have made comebacks in response to the banning of DDT in the US. Ironically, many of the pesticides which replaced DDT are more dangerous to humans, and, without DDT, disease (primarily in the tropics) claims more human lives.

HUMAN VS. NATURAL FOOD CHAINS

Human civilization is dependent on agriculture. Only with agriculture can a few people feed the rest of the population; the part of the population freed from raising food can then go on to do all the things we associate with civilization. Agriculture means manipulating the environment to favour plant species that we can eat. In essence, humans manipulate competition, allowing favoured species (crops) to thrive and thwarting species which might otherwise crowd them out (weeds). In essence, with agriculture we are creating a very simple ecosystem. At most, it has only three levels-producers (crops), primary consumers (livestock, humans) and secondary consumers (humans). This means that little energy is lost between tropic levels, since there are fewer trophic levels present.

This is good for humans, but what type of "ecosystem" have we created? Agricultural ecosystems have several problems. First, we create *monocultures*, or fields with only one crop. This is simplest for planting, weeding, and harvesting, but it also packs many similar plants into a small area, creating a situation ideal for disease and insect pests. In natural ecosystems, plants of one species are often scattered. Insects, which often specialize on feeding on a particular plant species, have a hard time finding the scattered plants. Without food, the insect populations are kept in check. In a field of corn however, even the most inept insect can find a new host plant with a jump in any direction. Likewise, disease is more easily spread if the plants are in close proximity. It takes lots of chemicals (pesticides) to keep a monoculture going.

Another problem with human agriculture is that we rely on relatively few plants for food. If the corn and rice crops failed worldwide in the same year, we would be hard-pressed to feed eve ryone. tural ecosystems usually have alternate sources of food available if one fails.

A final problem associated with agroecosystems is the problem of inorganic nutrient recycling. In a natural ecosystem, when a plant dies it fall to the ground and rots, and its inorganic nutrients are returned to the soil from which they were taken.

In human agriculture, however, we harvest the crop, truck it away, and flush it down the toilet to be run off in the rivers to the ocean. Aside from the water pollution problems this causes, it should be obvious to you that the nutrients are not returned to the fields. They have to be replaced with chemical fertilizers, and that means mining, transportation, electricity, etc. Also, the chemical fertilizers tend to run off the fields (along with soil disrupted by cultivation) and further pollute the water.

Some solutions are at hand, but they bring on new problems, too. No-till farming uses herbicides to kill plants in a field; the crop is then planted through the dead plants without plowing up the soil. This reduced soil and fertilizer erosion, but the herbicides themselves may damage ecosystems. In many areas, sewage sludge is returned to fields to act as a fertilizer. This reduces the need for chemical fertilizers, but still requires a lot of energy to haul the sludge around. Further, if one is not careful, things such as household chemicals and heavy metals may contaminate the sewage sludge and biomagnify in the crops which we would then eat.

BIOGEOCHEMICAL CYCLES

We have already seen that while energy does not cycle through an ecosystem, chemicals do. The inorganic nutrients cycle through more than the organisms, however, they also enter into the atmosphere, the oceans, and even rocks. Since these *chem*icals cycle through both the *bio*logical and the *geo*logical world, we call the overall cycles biogeochemical cycles. Each chemical has its own unique cycle, but all of the cycles do have some things in common. *Reservoirs* are those parts of the cycle where the chemical is held in large quantities for long periods of time. In *exchange pools*, on the other hand, the chemical is held for only a short time.

The length of time a chemical is held in an exchange pool or a reservoir is termed its *residence* time. The oceans are a reservoir for water, while a cloud is an exchange pool. Water may reside in an ocean for thousands of years, but in a cloud for a few days at best. The biotic community includes all living organisms. This community may serve as an exchange pool (although for some chemicals like carbon, bound in a sequoia for a thousand years, it may seem more like a reservoir), and also serve to move chemicals from one stage of the cycle to another.

For instance, the trees of the tropical rain forest bring water up from the forest floor to be evapourated into the atmosphere. Likewise, coral endosymbionts take carbon from the water and turn it into limestone rock. The energy for most of the transportation of chemicals from one place to another is provided either by the sun or by the heat released from the mantle and core of the Earth. While all inorganic nutrients cycle, we will focus on only 4 of the most important cycles-water, carbon (and oxygen), nitrogen, and phosphorous.

The Water Cycle

In the water cycle, energy is supplied by the sun, which drives evapouration whether it be from ocean surfaces or from treetops. The sun also provides the energy which drives the weather systems which move the water vapour (clouds) from one place to another (otherwise, it would only rain over the oceans). Precipitation occurs when water condenses from a gaseous state in the atmosphere and falls to earth. Evapouration is the reverse process in which liquid water becomes gaseous.

Gravity continues to operate, either pulling the water underground (groundwater) or across the surface (run-off). In either event, gravity continues to pull water lower and lower until it reaches the oceans (in most cases; the Great Salt Lake, Dead Sea, Caspian Sea, and other such depressions may also serve as the lowest basin into which water can be drawn). Frozen water may be trapped in cooler regions of the Earth (the poles, glaciers on mountaintops, etc.) as snow or ice, and may remain as such for very long periods of time. Lakes, ponds, and wetlands form where water is temporarily trapped. The oceans are salty because any weathering of minerals that occurs as the water runs to the ocean will add to the mineral content of the water, but water cannot leave the oceans except by evapouration, and evapouration leaves the minerals behind. Thus, rainfall and snowfall are comprised of relatively clean water, with the exception of pollutants (such as acids) picked up as the waster falls through the atmosphere. Organisms play an important role in the water cycle.

As you know, most organisms contain a significant amount of water. This water is not held for any length of time and moves out of the organism rather quickly in most cases. Animals and plants lose water through evapouration from the body surfaces, and through evapouration from the gas exchange structures (such as lungs). In plants, water is drawn in at the roots and moves to the gas exchange organs, the leaves, where it evapourates quickly. This special case is called transpiration because it is responsible for so much of the water that enters the atmosphere. In both plants and animals, the breakdown of carbohydrates (sugars) to produce energy (respiration) produces both carbon dioxide and water as waste products. Photosynthesis reverses this reaction, and water and carbon dioxide are combined to form carbohydrates. Now you understand the relevance of the term carbohydrate; it refers to the combination of carbon and water in the sugars we call carbohydrates.

Carbon Cycle

Once you understand the water cycle, the carbon cycle is relatively simple. From a biological perspective, the key events here are the complementary reactions of respiration and photosynthesis. Respiration takes carbohydrates and oxygen and combines them to produce carbon dioxide, water, and energy. Photosynthesis takes carbon dioxide and water and produces carbohydrates and oxygen. Photosynthesis takes energy from the sun and stores it in the carbon-carbon bonds of carbohydrates; respiration releases that energy. Both plants and animals carry on respiration, but only plants (and other producers) can carry on photosynthesis. The chief reservoirs for carbon dioxide are in the oceans and in rock.

Carbon dioxide dissolves readily in water. Once there, it may precipitate (fall out of solution) as a solid rock known as calcium carbonate (limestone). Corals and algae encourage this reaction and build up limestone reefs in the process. On land and in the water, plants take up carbon dioxide and convert it into carbohydrates through photosynthesis. This carbon in the plants now has 3 possible fates. It can be liberated to the atmosphere by the plant through respiration; it can be eaten by an animal, or it can be present in the plant when the plant dies. Animals obtain all their carbon in their food, and, thus, all carbon in biological systems ultimately comes from plants (autotrophs).

In the animal, the carbon also has the same 3 possible fates. Carbon from plants or animals that is released to the atmosphere through respiration will either be taken up by a plant in photosynthesis or dissolved in the oceans. When an animal or a plant dies, 2 things can happen to the carbon in it. It can either be respired by decomposers (and released to the atmosphere), or it can be buried intact and ultimately form coal, oil, or natural gas (fossil fuels). The fossil fuels can be mined and burned in the future; releasing carbon dioxide to the atmosphere.

Otherwise, the carbon in limestone or other sediments can only be released to the atmosphere when they are subducted and brought to volcanoes, or when they are pushed to the surface and slowly weathered away. Humans have a great impact on the carbon cycle because when we burn fossil fuels we release excess carbon dioxide into the atmosphere.

This means that more carbon dioxide goes into the oceans, and more is present in the atmosphere. The latter condition causes global warming, because the carbon dioxide in the atmosphere allows more energy to reach the Earth from the sun than it allows to escape from the Earth into space.

The Oxygen Cycle

If you look back at the carbon cycle, we have also described the oxygen cycle, since these atoms often are combined. Oxygen is present in the carbon dioxide, in the carbohydrates, in water, and as a molecule of two oxygen atoms. Oxygen is released to the atmosphere by autotrophs during photosynthesis and taken up by both autotrophs and heterotrophs during respiration. In fact, all of the oxygen in the atmosphere is *biogenic*; that is, it was released from water through photosynthesis by autotrophs. It took about 2 billion years for autotrophs (mostly cyanobacteria) to raise the oxygen content of the atmosphere to the 21% that it is today; this opened the door for complex organisms such as multicellular animals, which need a lot of oxygen.

The Nitrogen Cycle

The nitrogen cycle is one of the most difficult of the cycles to learn, simply because there are so many important forms of nitrogen, and because organisms are responsible for each of the interconversions. Remember that nitrogen is critically important in forming the amino portions of the amino acids which in turn form the proteins of your body. Proteins make up skin and muscle, among other important structural portions of your body, and all enzymes are proteins. Since enzymes carry out almost all of the chemical reactions in your body, it's easy how important nitrogen is. The chief reservoir of nitrogen is the atmosphere, which is about 78% nitrogen.

It is a pretty non-reactive gas; it takes a lot of energy to get nitrogen gas to break up and combine with other things, such as carbon or oxygen. Nitrogen gas can be taken from the atmosphere (fixed) in two basic ways. First, lightning provides enough energy to "burn" the nitrogen and fix it in the form of nitrate, which is a nitrogen with three oxygens attached. This process is duplicated in fertilizer factories to produce nitrogen fertilizers. These nitrogen-fixing bacteria come in three forms: some are free-living in the soil; some form symbiotic, mutualistic associations with the roots of bean plants and other legumes (rhizobial bacteria); and the third form of nitrogen-fixing bacteria are the photosynthetic cyanobacteria (blue-green algae) which are found most commonly in water. All of these fix nitrogen, either in the form of nitrate or in the form of ammonia (nitrogen with 3 hydrogens attached).

Most plants can take up nitrate and convert it to amino acids. Animals acquire all of their amino acids when they eat plants (or other animals). When plants or animals die (or release waste) the nitrogen is returned to the soil. The usual form of nitrogen returned to the soil in animal wastes or in the output of the decomposers, is ammonia. Ammonia is rather toxic, but, fortunately there are nitrite bacteria in the soil and in the water which take up ammonia and convert it to nitrite, which is nitrogen with two oxygens. Nitrite is also somewhat toxic, but another type of bacteria, nitrate bacteria, take nitrite and convert it to nitrate, which can be taken up by plants to continue the cycle. We now have a cycle set up in the soil (or water), but what returns nitrogen to the air? It turns out that there are denitrifying bacteria which take the nitrate and combine the nitrogen back into nitrogen gas.

The nitrogen cycle has some important practical considerations, as anyone who has ever set up a saltwater fish tank has found out. It takes several weeks to set up such a tank, because you must have sufficient numbers of nitrite and nitrate bacteria present to detoxify the ammonia produced by the fish and decomposers in the tank. Otherwise, the ammonia levels in the tank will build up and kill the fish. This is usually not a problem in freshwater tanks for two reasons. One, the pH in a freshwater tank is at a different level than in a saltwater tank. At the pH of a freshwater tank, ammonia is not as toxic. Second, there are more multicellular plant forms that can grow in freshwater, and these plants remove the ammonia from the water very efficiently. It is hard to get enough plants growing in a saltwater tank to detoxify the water in the same way.

The Phosphorous Cycle

The phosphorous cycle is the simplest of the cycles. Phosphorous has only one form, phosphate, which is a phosphorous atom with 4 oxygen atoms. This heavy molecule never makes its way into the atmosphere, it is always part of an organism, dissolved in water, or in the form of rock. When rock with phosphate is exposed to water (especially water with a little acid in it), the rock is weathered out and goes into solution. Autotrophs take this phosphorous up and use it in a variety of ways. It is an important constituent of cell membranes, DNA, RNA, and, of course ATP, which, after all, stands for adenosine tri*phosphate*.

Heterotrophs (animals) obtain their phosphorous from the plants they eat, although one type of heterotroph, the fungi, excel at taking up phosphorous and may form mutualistic symbiotic relationships with plant roots. These relationships are called *mycorrhizae*; the plant gets phosphate from the fungus and gives the fungus sugars in return. Animals, by the way, may also use phosphorous as a component of bones, teeth and shells. When animals or plants die (or when animals defecate), the phosphate may be returned to the soil or water by the decomposers. There, it can be taken up by another plant and used again.

This cycle will occur over and over until at last the phosphorous is lost at the bottom of the deepest parts of the ocean, where it becomes part of the sedimentary rocks forming there. Ultimately, this phosphorous will be released if the rock is brought to the surface and weathered. Two types of animals play a unique role in the phosphorous cycle.

The phosphate is then used as fertilizer. This mining of phosphate and use of the phosphate as fertilizer greatly accelerates the phosphorous cycle and may cause local overabundance of phosphorous, particularly in coastal regions, at the mouths of rivers, and anyplace where there is a lot of sewage released into the water (the phosphate placed on crops finds its way into our stomachs and from there to our toilets). Local abundance of phosphate can cause overgrowth of algae in the water; the algae can use up all the oxygen in the water and kill other aquatic life.

This is called eutrophication. The other animals that play a unique role in the phosphorous cycle are marine birds. These birds take phosphorous containing fish out of the ocean and return to land, where they defecate. Their guano contains high levels of phosphorous and in this way marine birds return phosphorous from the ocean to the land. The guano is often mined and may form the basis of the economy in some areas.

Energy Transformations and Biogeochemical Cycling

The study of ecosystems mainly consists of the study of certain processes that link the living, or biotic, components to the non-living, or abiotic, components. *Energy transformations* and *biogeochemical cycling* are the main processes that comprise the field of ecosystem ecology. As we learned earlier, ecology generally is defined as the interactions of organisms with one another and with the environment in which they occur. We can study ecology at the level of the individual, the population, the community, and the ecosystem. Studies of *individuals* are concerned mostly about physiology, reproduction, development or behaviour, and studies of *populations* usually focus on the habitat and resource needs of individual species, their group behaviours, population growth, and what limits their abundance or causes extinction.

Studies of *communities* examine how populations of many species interact with one another, such as predators and their prey, or competitors that share common needs or resources. In *ecosystem ecology* we put all of this together and, insofar as we can, we try to understand how the system operates as a whole. This means that, rather than worrying mainly about particular species, we try to focus on major functional aspects of the system.

These *functional aspects* include such things as the amount of energy that is produced by photosynthesis, how energy or materials flow along the many steps in a food chain, or what controls the rate of decomposition of materials or the rate at which nutrients are recycled in the system.

Components of an Ecosystem

A basic understanding of the diversity of plants and animals, and how plants and animals and microbes obtain water, nutrients, and food. We can clarify the parts of an ecosystem by listing them under the headings "abiotic" and "biotic".

Abiotic Components	Biotic Components
Sunlight	Primary producers
Temperature	Herbivores
Precipitation	Carnivores
Water or moisture	Omnivores
Soil or water chemistry (*e.g.*, P, NH_4^+)	Detritivores

By and large, this set of environmental factors is important almost everywhere, in all ecosystems. A *functional group* is a biological category composed of organisms that perform mostly the same kind of function in the system; all the photosynthetic plants or primary producers form a functional group. Membership in the functional group does not depend very much on who the actual players happen to be, only on what function they perform in the ecosystem.

Ecosystems

This figure with the plants, zebra, lion, and so forth illustrates the two main ideas about how ecosystems function: ecosystems have energy flows and ecosystems cycle materials.

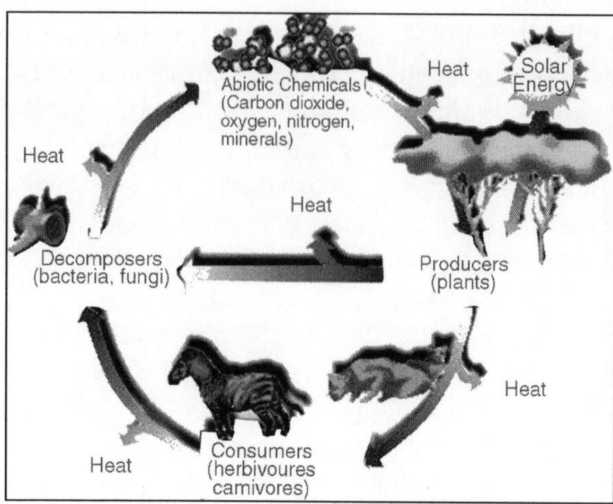

Fig. Energy Flows and Material Cycles.

These two processes are linked, but they are not quite the same energy enters the biological system as light energy, or photons, is transformed into chemical energy in organic molecules by cellular processes including photosynthesis and respiration, and ultimately is converted to heat energy. This energy is dissipated, meaning it is lost to the system as heat; once it is lost it cannot be recycled. Without the continued input of solar energy, biological systems would quickly shut down. Thus the earth is an *open system* with respect to energy. Elements such as carbon, nitrogen, or phosphorus enter living organisms in a variety of ways. Plants obtain elements from the surrounding atmosphere, water, or soils. Animals may also obtain elements directly from the physical environment, but usually they obtain these mainly as a consequence of consuming other organisms.

These materials are transformed biochemically within the bodies of organisms, but sooner or later, due to excretion or decomposition, they are returned to an inorganic state. Often bacteria complete this process, through the process called decomposition or mineralization During decomposition these materials are not destroyed or lost, so the earth is a *closed system* with respect to elements (with the exception of a meteorite entering the system now and then). The elements are cycled endlessly between their biotic and abiotic states within ecosystems. Those elements whose supply tends to limit biological activity are called *nutrients*.

The Transformation of Energy

The transformations of energy in an ecosystem begin first with the input of energy from the sun. Energy from the sun is captured by the process of photosynthesis. Carbon dioxide is combined with hydrogen (derived from the splitting of water molecules) to produce carbohydrates (CHO). Energy is stored in the high energy bonds of adenosine triphosphate.

The prophet Isaah said "all flesh is grass", earning him the title of first ecologist, because virtually all energy available to organisms originates in plants. Because it is the first step in the production of energy for living things, it is called *primary production Herbivores* obtain their energy by consuming plants or plant products, *carnivores* eat herbivores, and *detritivores* consume the droppings and carcasses of us all. portrays a simple food chain, in which energy from the sun, captured by plant photosynthesis, flows from *trophic level* to trophic level via the *food chain*.

A trophic level is composed of organisms that make a living in the same way, that is they are all *primary producers* (plants), *primary consumers* (herbivores) or *secondary consumers* (carnivores). Dead tissue and waste products are produced at all levels.

Scavengers, detritivores, and decomposers collectively account for the use of all such "waste"—consumers of carcasses and fallen leaves may be other animals, such as crows and beetles, but ultimately it is the microbes that finish the job of decomposition. Not surprisingly, the amount of primary production varies a great deal from place to place, due to differences in the amount of solar radiation and the availability of nutrients and water. Energy transfer through the food chain is inefficient. This means that less energy is available at the herbivore level than at the primary producer level, less yet at the carnivore level, and so on. The result is a pyramid of energy, with important implications for understanding the quantity of life that can be supported. Usually when we think of food chains we visualize green plants, herbivores, and so on. These are referred to as *grazer food chains*, because living plants are directly consumed. In many circumstances the principal energy input is not green plants but dead organic matter.

These are called *detritus food chains*. Examples include the forest floor or a woodland stream in a forested area, a salt marsh, and most obviously, the ocean floor in very deep areas where all sunlight is extinguished 1000's of meters above. In subsequent lectures we shall return to these important issues concerning energy flow. Finally, although we have been talking about food chains, in reality the organization of biological systems is much more complicated than can be represented by a simple "chain". There are many food links and chains in an ecosystem, and we refer to all of these linkages as a *food web*. Food webs can be very complicated, where it appears that *"everything is connected to everything else"*, and it is important to understand what are the most important linkages in any particular food web.

BIOGEOCHEMISTRY

One obvious way is to study the flow of energy or the cycling of elements. The cycling of elements is controlled in part by organisms, which store or transform elements, and in part by the chemistry and geology of the natural world. The term *Biogeochemistry* is defined as the study of how living systems influence, and are controlled by, the geology and chemistry of the earth. Thus biogeochemistry encompasses many aspects of the abiotic and biotic world that we live in.

There are several main *principles and tools* that biogeochemists use to study earth systems. Most of the major environmental problems that we face in our world toady can be analysed using biogeochemical principles and tools. These problems include global warming, acid rain, environmental pollution, and increasing greenhouse gases. The principles and tools that we use can be broken down into 3 major components: *element ratios, mass balance, and element cycling*.

Element Ratios

In biological systems, we refer to important elements as *"conservative"*. These elements are often nutrients. By "conservative" we mean that an organism can change only slightly the amount of these elements in their tissues if they are to remain in good health. It is easiest to think of these conservative elements in relation to other important elements in the organism. In healthy algae the elements C, N, P, and Fe have the following ratio, called the *Redfield ratio* after the oceanographer who discovered it:

C: N: P: Fe = 106: 16: 1: 0.01

Once we know these ratios, we can compare them to the ratios that we measure in a sample of algae to determine if the algae are lacking in one of these limiting nutrients.

Mass Balance

Another important tool that biogeochemists use is a simple mass balance equation to describe the state of a system. The system could be a snake, a tree, a lake, or the entire globe. Using a mass balance approach we can determine whether the system is changing and how fast it is changing.

The equation is:

Net Change = Input + Output + Internal Change

In this equation the net change in the system from one time period to another is determined by what the inputs are, what the outputs are, and what the internal change in the system was. The example given in class is of the acidification of a lake, considering the inputs and outputs and internal change of acid in the lake.

Element Cycling

Element cycling describes where and how fast elements move in a system. There are two general classes of systems that we can analyse. A closed system refers to a system where the inputs and outputs are negligible compared to the internal changes. Examples of such systems would include a bottle, or our entire globe. There are two ways we can describe the cycling of materials within this closed system, either by looking at the rate of movement or at the pathways of movement.

- Rate = number of cycles/time * as rate increases, productivity increases
- Pathways-important because of different reactions that may occur

In an open system there are inputs and outputs as well as the internal cycling. Thus we can describe the rates of movement and the pathways, just as we did for the closed system, but we can also define a new concept called the residence time. The residence time indicates how long on average an element remains within the system before leaving the system.

- Rate
- Pathways
- Residence time, Rt
- Rt = Total amount of matter/output rate of matter

Controls on Ecosystem Function

Now that we have learned something about how ecosystems are put together and how materials and energy flow through ecosystems, we can better address the question of "what controls ecosystem function"? There are two dominant theories of the control of ecosystems. The first, called bottom-up control, states that it is the nutrient supply to the primary producers that ultimately controls how ecosystems function.

If the nutrient supply is increased, the resulting increase in production of autotrophs is propagated through the food web and all of the other trophic levels will respond to the increased availability of food (energy and materials will cycle faster). The second theory, called top-down control, states that predation and grazing by higher trophic levels on lower trophic levels ultimately controls ecosystem function. If you have an increase in predators, that increase will result in fewer grazers, and that decrease in grazers will result in turn in more primary producers because fewer of them are being eaten by the grazers. Thus the control of population numbers and overall productivity "cascades" from the top levels of the food chain down to the bottom trophic levels. So, which theory is correct? Well, as is often the case when there is a clear dichotomy to choose from, the answer lies somewhere in the middle.

There is evidence from many ecosystem studies that both controls are operating to some degree, but that neither control is complete. The "top-down" effect is often very strong at trophic levels near to the top predators, but the control weakens as you move further down the food chain. Similarly, the "bottom-up" effect of adding nutrients usually stimulates primary production, but the stimulation of secondary production further up the food chain is less strong or is absent. Thus we find that both of these controls are operating in any system at any time, and we must understand the relative importance of each control in order to help us to predict how an ecosystem will behave or change under different circumstances, such as in the face of a changing climate.

The Geography of Ecosystems

There are many different ecosystems: rain forests and tundra, coral reefs and ponds, grasslands and deserts. Climate differences from place to place largely determine the types of ecosystems. How terrestrial ecosystems appear to us is influenced mainly by the dominant vegetation. The word "biome" is used to describe a major vegetation type such as tropical rain forest, grassland, tundra, etc., extending over a large geographic area. It is never used for aquatic systems, such as ponds or coral reefs. It always refers to a vegetation category that is dominant over a very large geographic scale, and so is somewhat broader than an ecosystem. Every place on earth gets the same total number of hours of sunlight each year, but not the same amount of heat.

The sun's rays strike low latitudes directly but high latitudes obliquely. This uneven distribution of heat sets up not just temperature differences, but global wind and ocean currents that in turn have a great deal to do with where rainfall occurs. Add in the cooling effects of elevation and the effects of land masses on temperature and rainfall, and we get a complicated global pattern of climate.

A schematic view of the earth shows that, complicated though climate may be, many aspects are predictable High solar energy striking near the equator ensures nearly constant high temperatures and high rates of evapouration and plant transpiration. Warm air rises, cools, and sheds its moisture, creating just the conditions for a tropical rain forest. Contrast the stable temperature but varying rainfall of a site in Panama with the relatively constant precipitation but seasonally changing temperature of a site in New York State. Every location has a rainfall- temperature graph that is typical of a broader region.

We can draw upon plant physiology to know that certain plants are distinctive of certain climates, creating the vegetation appearance that we call biomes. Note how well the distribution of biomes plots on the distribution of climates Note also that some climates are impossible, at least on our planet. High precipitation is not possible at low temperatures—there is not enough solar energy to power the water cycle, and most water is frozen and thus biologically unavailable throughout the year.

The high tundra is as much a desert as is the Sahara:

- Ecosystems are made up of abiotic (non-living, environmental) and biotic components, and these basic components are important to nearly all types of ecosystems. Ecosystem Ecology looks at energy transformations and biogeochemical cycling within ecosystems.
- Energy is continually input into an ecosystem in the form of light

energy, and some energy is lost with each transfer to a higher trophic level. Nutrients, on the other hand, are recycled within an ecosystem, and their supply normally limits biological activity. So, "energy flows, elements cycle".

- Energy is moved through an ecosystem via a food web, which is made up of interlocking food chains. Energy is first captured by photosynthesis (primary production). The amount of primary production determines the amount of energy available to higher trophic levels.
- The study of how chemical elements cycle through an ecosystem is termed biogeochemistry. A biogeochemical cycle can be expressed as a set of stores (pools) and transfers, and can be studied using the concepts of "stoichiometry", "mass balance", and "residence time".
- Ecosystem function is controlled mainly by two processes, "top-down" and "bottom-up" controls.
- A biome is a major vegetation type extending over a large area. Biome distributions are determined largely by temperature and precipitation patterns on the Earth's surface.

BIOGEOCHEMISTRY

As the terminology implies, Biogeochemistry is a borderline specialisation involving three basic branches of natural, physical and earth sciences. On one hand living organisms are involved and on the other substances such as nutrients, are made available to the living organisms either through soil (the pedosphere) or through water (the hydrosphere). Of course there are certain organisms which can get their food through air (atmosphere) also.

The living organisms, while differing vastly among themselves in size, morphology and physiology, share, all the same a basic feature common to them all: ability to metabolise their requirements in the habitats they live. An individual organism is perishable but life process, sustained by successive generation, is perpetual. Biogeochemistry is all about impact of such living processes on our environment. The bulk of living matter is concentrated in the landmass of the world though at the continent-ocean interface and at the bottom of the sea they are present in abundance. The physio-chemical interaction between organisms and the land that sustain them have been a subject of study for over a century now.

More than two hundred years ago, The turnover of elements on the surface of the globe in which he postulated on the cyclic exchange of elements among the three kingdoms of nature, namely mineral, vegetable and animal. Lavoisier paid a heavy price for stating so: he was executed after his work was published. The interaction among the above three kingdoms essentially constitute the broad field of Biogeochemistry as we know today.

The original kingdoms have been replaced by reservoirs of different kind and atmosphere is brought into the picture because of gases involved in many biogeochemical processes. Since life is cyclic, all processes where living matter is involved has also got to be cyclic. Hence, in all the reservoirs are linked to each other in both directions:at the inlet and outlet point for each reservoir. The interconnection of all the reservoirs leads to a overall global cycle. Such a cycle is known by various names depending on the user and the sector of interest: biogeochemical cycle, hydrological cycle, geochemical cycle and rock cycle. The emphasis in each cycle varies: in hydrological cycle, one will be interested in transfer of water; in the geochemical cycle, one will be interested in transfer from the crust to other segments and in rock cycle, the interest will be transformation of large volume of crustal materials across various interfaces. Whereas in the biogeochemical cycle the focus is on nutrients essential for the life processes and at the toxic substances determinant to such living things.

Thus biogeochemical processes can be defined in three stages:

- *Stage*-1: Elements to be grouped according to the needs:
  Essential, such as Carbon (C), Nitrogen (N) and Phosphorous (P)
  Minor, such as Sodium (Na), Potassium(K) etc.
  Trace, such as Iron (Fe), Manganese (Mn), Copper (Cu) etc.
  Toxic, such as Mercury (Hg), Selenium (Se), Cadmium (Cd) (all the above groupings according to the requirements of biological system in the biological world)-the bio part of the topic.
- *Stage*-2: The availability of the above from the earth, generally from the top few metres of the earth (the geocomponent, more specifically the pedosphere the geo part of the topic and now.
- *Stage*-3: The physio-chemical mechanisms that make the needs of the biological systems from the pedospheric layers of the earth in response to micro-environmental functions such as pH (acidity), weathering and release of elements by their breakdown due to chemical interaction of individual minerals with water. This constitutes the chemical part of the topic.

Let this linkages using an essential element Carbon as an example. Referring to carbon is present in the atmosphere mainly as carbon dioxide (CO_2) and in small quantities as carbon monoxides (CO). In the hydrosphere, carbon is present as bicarbonate and carbonate the major factor in the hardness of water. In the lithosphere- pedosphere, it is present as calcium carbonate ($CaCO_3$) or some combination thereof. In the biosphere, carbon is present as the building block of the plants and animals in the form of variety of carbon compounds such as carbohydrates, proteins and amino-acids. Carbon enters the biological system either through geological process (weathering of carbon-bearing minerals such as limestone ($CaCO_3$) and/or chemical process directly from the atmosphere (CO_2) by photosynthesis.

When the organism (including plant) dies out, the carbon is released back either directly to the atmosphere as CO_2 or is mineralised as coal, limestone etc. So that it is returned to the lithosphere. Thus the carbon cycle is complete. The oldest carbon-bearing organic world remains known to man is about 3.5 billion years whereas the age of the earth is at present estimated to be about 4.5 billion years. In spite of the complexity of the biological evolution through geological time scale, carbon has been and is being cycled through various reservoirs relatively unaffected and un- disturbed till recent times.

Among the three essential nutrients for life phosphorous (P) is extremely interesting. Unlike carbon and nitrogen, it is predominately derived from the crustal top of the solid earth and mobilised through the biosphere by the transport behaviour in the hydrosphereP ignites in air but glows when left in darkness. Thus it is a carrier of energy in all living systems. The homologous elements P and N has complimentary properties in relation to Hydrogen(H) and Oxygen(O_2)- the other important elements for most living matter on the surface of the earth. P in association with O_2 introduces some structural order in the molecules at cellular levels and thus the P-O association in the form of PO_4 units separates life from ordinary inanimate minerals represented by similar association of Silicon with O_2 (SiO_4)

From a single cell amoeba some 3.5 billion years ago (3500 million), organisms progressively went through complex biological evolution as a function of geological time scale. The link between time and evolution is established through changing environmental parameters such the carbon dioxide (CO_2) levels in atmosphere prevailing at a given time, the O_2 levels, nutrients such as P and minor and trace elements etc. Geochemical evidences through rock records show that about 2.5 billion years ago, the atmospheric CO_2 was at least 100 times lower as that of today and O_2 levels negligible compared to today s environment.

Have been continuously interacting among each other through such a vast time scale to maintain the cyclic behaviour of nutrients and other elements. Excessive constituents in any one segment was removed over a period of time to other segments so that its biogeochemical cyclic behaviour was not affected. During Gondwana times (about 400 to 500 million years before) when India, Australia and Antarctica were neighbours, excess CO_2 in the atmosphere was neutralised by enhanced photosynthesis and subsequently the biota decayed and gave raise to most of the coal deposits which we are exploiting today. Thus due to the biogeochemical behaviour, nature was able to adjust itself at its own rate in any parameters. On the other hand in modern times, man introduces any one variable to a segment in very abruptly and in such large dose that nature is unable to adjust to that change. A case in point the emission of green house gas CO_2 by the burning of coal. The essential nutrient element C is transferred from the lithosphere to the atmosphere so fast that the excess CO_2 is unable to be removed equally faster by the hydrosphere leading to enhanced levels of CO_2 in the atmosphere and hence we hear everyday about the global warming On the other hand, let us take a element such as Mercury (Hg) considered to be very toxic to the biological world. The movement of Hg in cyclic fashion through our environment. Hg generally occurs in minerals in association with sulphur in the form of mercury sulphide (HgS) systems.

Even in rocks, it is considered to be trace element since its abundance in the crustal earth is very low. However, during the exploitation of base metals resources, such as that of copper, lead and Zinc, mercury is extracted as a buy-product. Modern civilization uses mercury for a variety of products such as chemicals, coatings on seeds etc. Even though it does not react with water in the hydrosphere easily, mercury combines with any organic substances such decayed plant debris etc and immediately becomes toxic to the living biota. The well known minemata decease in Japan was the result of such affinity of mercury for many biotic materials. Being the only metal known to man to be in liquid state at room conditions, it easily vapourises and hence get also into the atmospheric segment Hence due to man s use, what was once locked up safely in inert geological materials in rocks, have been mobilised to the biosphere, hydrosphere and atmosphere.

The story is basically same for a few other toxic metals such as Cadmium though the details may individually vary due to their differing biogeochemical behaviour in our present environment.

It is clear that processes in our environment are better understood by focussing on the interfacing areas. Thus Biogeochemistry is the ideal such frontier area of study and research. It is however a relatively new field and only in the last two decades various researchers across the world have been focussing their attention on this inter- disciplinary subject. Outside India there are a few places where separate institutes exists for this purpose. At about the same time the Ministry of Environment and Forests also approved the setting up of a ENVIS centre in JNU on the same subject under his direction. Though there no other specific institutions in India in this subject, there are individual researchers in many universities working on topics of interest to Biogeochemistry.

Thus, the area of Biogeochemistry is an emerging branch of multi-diciplinary approach involving all major branches of Sciences. From a humble beginning two decades ago, now this subject is well established in the international field and in India is fast catching up as an important tool of research for understanding processes in our environment. Sufficient infra- structure has now developed in India through a variety of individual and institutional efforts so that in the years to come the importance work on Biogeochemistry will be better understood and appreciated. The involvement of a number of specialists is obvious when ones looks into the above three stages: for understanding the biotic behaviour, one should have a good understanding of ecological principles; the requirements for the ecosystem is provided by basic geochemical processes of weathering helped by microbiological and biochemical reactions in soils and over rocks; farther, from the source to the receptor, chemical and hydrological principles operate as a middleman. Hence within the broad field of Biogeochemistry itself, there are sub-specialisations depending on the background of the researchers- be a ecologist, geochemist or chemical hydrologist. Hence the word Biogeochemistry is truly an integrating field to better understand our present, past and future environmental problems.

Among the more recent sub-specialisations within Biogeochemistry that have emerged in the international research is the application of certain biochemical and molecular biochemical techniques to understand our present and past environment. A full human body was found in tact in the frozen alps and several investigations on the body using most modern biochemical techniques have thrown some light about the habitat of the community some 5000 years ago, then the prevailing eco- system, palaeo-environment, anthropological lineage etc.

DNA based study thus offers excellent scope even in environment related matter. In India, a student from our labouratory is working under the supervision of Prof. Laljit Singh, Centre for DNA Finger-printing at Hyderabad to apply DNA techniques in fossils bones of geological and anthropological importance to understand the habitat and palaeo-environment. DNA techniques to ancient biological specimens. Farther, characterisations of amino-acids in ancient sediments have been successfully used to understand palaeo-environment and also stresses acting on such systems.

Thus, Biogeochemistry can be viewed as a very challenging area of study involving a variety of basic sciences so that it is ideally suited to understand our environment better. Research should not go in the direction of reinventing the wheel but improve and if possible, replace it. Hopefully the 21st century consolidation of the field of Biogeochemistry as an established brach of science. Enlarged version of specific component. whenevr possible. Important components in each segment and direction of transfer F means flux that is concentration for a given parameter the volume of water/air transferred in that direction per unit time. This, F has unit of wt/vol/time. F is likely to have volumes since the time involved is relatively shorter for some of the sectors. For eg. some of the variables transferred involving soil and plants/animals and plants and atmosphere etc. may have a time scale of a few seconds only.

The carbon transfer, the following fluxes have been incorporated: 1 to 34 point all have same magnitude, time scale and importance. Between river and young soils sediments, the transfer 26 and 27 represent deposition and erosion processes. Faster compared to deposition or accumulation so that net transfer is loss of carbon from soil (Pedosphere) to river (hydrosphere). Hg cycle is shwn in tow stages: pre-man and present day. this concept is adapted from Garrels *et al.,* 1976. The main differences is indicated in teh Hg-mining in the present day cause additional flux-vaporisation-from land to atmosphere. + + or + singns indicate higher gains in the respectinve sectors the present day cycle relative to pre-man cycle. As a consequence there is also poportional increase in the flxes in the present day cycle between the concerned sectors. Garels et al. (1976) estimate that Hg will be the first metal that is likely to be lost out entiely ot of the solid spheres (litho-and pedospheres) to the fluid (hydro- and atmo-) spheres in foreseabe future.

Mercury in the Environment

Mercury is a silvery, liquid metal at room temperature and is often referred to as one of the "heavy metals." Like water, mercury can evapourate and become airborne. Because it is an element, mercury does not break down into less toxic substances. Once mercury escapes to the environment, it circulates in and out of the atmosphere until it ends up in the bottoms of lakes and oceans.

Mercury can be found as the elemental metal or in a wide variety of organic and inorganic compounds. Depending on its chemical form, mercury may travel long distances before it falls to earth with precipitation or dust. Bacteria and chemical reactions in lakes and wetlands can change the mercury into a much more toxic form known as methylmercury. Fish become contaminated with methylmercury by eating food (plankton and smaller fish), which has absorbed methylmercury.

As long as the fish continue to be exposed to mercury, mercury continually builds up in fish's bodies. Fish that eat other fish become even more highly contaminated. Thus, the largest tend to be the most contaminated. When people eat the contaminated fish, the methylmercury can remain in their bodies for a long time. If they eat fish containing methylmercury faster than their bodies can get discharge it, the methylmercury accumulates in their bodies and can be toxic. Many states have fish consumption advisories to inform people about how many meals of fish they can safely eat over a period of time.

Mercury's Environmental Effects

Fish are the main source of food for many birds and other animals, and mercury can seriously damage the health of these species. Loons, eagles, panthers, otters, mink, kingfishers and ospreys naturally eat large quantities of fish. Because these predators rely on speed and coordination to obtain food, mercury may be particularly hazardous to these animals.

Recent research in Minnesota indicates that the following environmental effects are occurring:

- Loons are accumulating so much mercury that it may be affecting their ability to reproduce;

- Elevated levels of mercury have been found in mink and otters;
- Walleye reproduction may be impaired by the fish's exposure to mercury.

Similar effects are being documented for other fish and fish-eating species around the United States and Canada. Has there always been mercury contamination, or is this a recent problem? This is a difficult question to answer, in part because of a lack of adequately preserved fish specimens of preindustrial age to compare against contemporary samples. However, several lines of evidence from recent studies on Wisconsin lakes suggest that increased emissions to the atmosphere, and subsequent higher deposition rates to lakes, likely translate into higher mercury levels in fish.

The Mercury Cycle and Bioaccumulation

There is a constant biogeochemical cycle of mercury. This cycle includes:

- Release of elemental mercury as a gas from the rocks and waters (degassing);
- Long-range transport of the gases in the atmosphere;
- Wet and dry deposition upon land and surface water;
- Absorption onto sediment particles;
- Bioaccumulation (or biomagnification) in terrestrial and aquatic food chains.

Bioaccumulation means an increase in the concentration of a chemical in an organism over time, compared to the chemical's concentration in the environment. Bioaccumulation can be a normal and essential process for the growth of any species, but the accumulation of unnecessary chemicals or toxins, or even the overaccumulation of essential substances can be detrimental. All animals, including humans, daily bioaccumulate many vital nutrients, such as vitamins A, D, and K, trace minerals, essential fats and amino acids, but unfortunately, they can also accumulate many unnecessary substances, such as lead or mercury. What concerns toxicologists is the bioaccumulation of necessary substances to levels in the body that can cause harm.

With substances such as lead or mercury, any accumulation at all can be harmful. Compounds accumulate in living things any time they are taken up and stored faster than they are broken down (metabolized) or excreted. Understanding the dynamic process of bioaccumulation is important in protecting humans and other organisms from the adverse effects of chemical exposure, and it has become a critical consideration in the regulation of chemicals. Bioaccumulation varies among individual organisms as well as among species. Large, fat, long-lived individuals or species with low rates of metabolism or excretion of a chemical will bioaccumulate more than small, thin, short-lived organisms. Thus, an old lake trout may bioaccumulate much more than a young bluegill in the same lake.

Above is a schematic drawing of mercury cycling in an aquatic ecosystem. With the exception of isolated cases of known point sources, the source of most mercury to most aquatic ecosystems is deposition from the atmosphere, primarily associated with rainfall.

In the aquatic environment, mercury can be:

- Dissolved or suspended in the water
- Trapped in the sediments
- Ingested by living things (biota)

Methylmercury is the form of mercury most available and most toxic to biota (including zooplankton, insects, fish, and humans). This form of mercury is easily taken up by biota and bioaccumulates in their tissues. Unlike many other fish contaminants, such as PCBs, dioxin, and DDT, mercury does not concentrate in the fat, but in the muscle tissue. Thus, there is no simple way to remove mercury-contaminated portions from fish that is to be eaten.

LIFE AND BIOGEOCHEMICAL CYCLES

THE SULFUR CYCLE

Another example of a major biogeochemical cycle of significance to climate and life is the sulfur cycle. Living things require certain safe, low levels of this nutrient. The sulfur cycle can be thought of as beginning with the gas sulfur dioxide (SO_2) or the particles of sulfate (SO_4=) compounds in the air. These compounds either fall out or are rained out of the atmosphere. Plants take up some forms of these compounds and incorporate them into their tissues. Then, as with nitrogen, these organic sulfur compounds are returned to the land or water after the plants die or are consumed by animals. Bacteria are important here as well since they can transform the organic sulfur to hydrogen sulfide gas (H_2S). In the oceans, certain phytoplankton can produce a chemical that transforms to SO_2 that resides in the atmosphere. These gases can re-enter the atmosphere, water, and soil, and continue the cycle.

In its reduced oxidation state, the nutrient sulfur plays an important part in the structure and function of proteins. In its fully oxidized state, sulfur exists as sulfate and is the major cause of acidity in both natural and polluted rainwater. This link to acidity makes sulfur important to geochemical, atmospheric, and biological processes such as the natural weathering of rocks, acid precipitation, and rates of denitrification. Sulfur is also one of the main elemental cycles most heavily perturbed by human activity. Estimates suggest that emissions of sulfur to the atmosphere from human activity are at least equal or probably larger in magnitude than those from natural processes. Like nitrogen, sulfur can exist in many forms: as gases or sulfuric acid particles. Sulfuric acid particles contribute to the polluting smog that engulfs some industrial centres and cities where many sulfur containing fuels are burned. Such particles floating in air (known as sulfate aerosols) can cause respiratory diseases or cool the climate by reflecting some extra sunlight to space.

The lifetime of most sulfur compounds in the air is relatively short (*e.g.* days). Superimposed on these fast cycles of sulfur are the extremely slow sedimentary-cycle processes or erosion, sedimentation, and uplift of rocks containing sulfur. In addition, sulfur compounds from volcanoes are intermittently injected into the atmosphere, and a continual stream of these compounds is produced from industrial activities. These compounds mix with water vapour and form sulfuric acid smog. In addition to contributing to acid rain, the sulfuric acid droplets of smog form a haze layer that reflects solar radiation and can cause a cooling of the earth's surface. While many questions remain concerning specifics, the sulfur cycle in general, and acid rain and smog issues in particular are becoming major physical, biological, and social problems.

BIOGEOCHEMICAL CYCLES

The Earth is a closed system for matter, except for small amounts of cosmic debris that enter the Earth's atmosphere. This means that all the elements needed for the structure and chemical processes of life come from the elements that were present in the Earth's crust when it was formed billions of years ago. This matter, the building blocks of life, continually cycle through Earth's systems, the atmosphere, hydrosphere, biosphere, and lithosphere, on time scales that range from a few days to millions of years. These cycles are called biogeochemical cycles, because they include a variety of biological, geological, and chemical processes. Many elements cycle through ecosystems, organisms, air, water, and soil. Many of these are trace elements. Other elements, including carbon, nitrogen, oxygen, hydrogen, sulfur, and phosphorus are critical components of all biological life. Together, oxygen and carbon account for 80 per cent of the weight of human beings. Because these elements are key components of life, they must be available for biological processes.

Carbon, however, is relatively rare in the Earth's crust, and nitrogen, though abundant in the atmosphere, is in a form that is not useable by living organisms.The biogeochemical cycles transport and store these important elements so that they can be used by living organisms. Each cycle takes many different pathways and has various reservoirs, or storage places, where elements may reside for short or long periods of time. Each of the chemical, biological, and geological processes varies in their rates of cycling. Some molecules may cycle very quickly depending on the pathway. Carbon atoms in deep ocean sediments may take hundreds to millions of years to cycle completely through the system. An average water molecule resides in the atmosphere for about ten days, although it may be transported many miles before it falls back to the Earth as rain. How fast substances cycle depends on its chemical reactivity and whether or not it can be found in a gaseous state. A gaseous phase allows molecules to be transported quickly. Phosphorous has no gaseous phase and is relatively uncreative, so it moves very slowly through its cycle. Phosphorus is stored in large amounts in sediment in the oceans or in the Earth's crust and is recycled back to the surface only over very long periods of time through upwelling of ocean waters or weathering of rocks.

Biogeochemical cycles are subject to disturbance by human activities. Humans accelerate natural biogeochemical cycles when elements are extracted from their reservoirs, or sources, and deposited back into the environment (sinks). Humans have significantly altered the carbon cycle by extracting and combusting billions of tons of hydrocarbons in fossil that were buried deep in the Earth's crust, in addition to clearing vegetation that stores carbon. Global release of carbon through human activities has increased from 1 billion tons per year in 1940 to 6.5 billion tons per year in 2000.

About half of this extra carbon is taken up by plants and the oceans, while the other half remains in the atmosphere. In addition to carbon cycle, humans have altered the nitrogen and phosphorus cycles by adding these elements to croplands as fertilizers, which has contributed to over-fertilization of aquatic ecosystems when excess amounts are carried by run-off into local waterways. Researchers are trying to understand all of the various pathways and flows of each of the biogeochemical cycles in order to understand how human activities affect these cycles. While many important processes have been understood for more than century, there are many phenomena that scientists are just beginning to investigate. Satellite technology, among other tools, has revealed new information about interactions between the oceans and atmosphere that contribute to knowledge about the carbon cycle, but there remain many unanswered questions.

Human Modifications of Climate Services

Human activities are significantly perturbing all of these biogeochemical cycles as well as other earth system processes, both directly through industrial processes and indirectly through changing distributions and abundance of life. The atmosphere is of particular importance to the perturbations due to its crucial role in mediating all energy that enters and leaves earth. Overall, the atmosphere is the component that controls the dominant energy flow in the earth's climate system, and solar radiation from the sun provides the energy to make the weather machine work.

Embedded in this process are the biogeochemical cycles we have described that operate on a variety of time and space scales and help to regulate flows of energy and materials throughout the earth system Yet, while we understand much about the functioning of separate parts of this system, there is still a great deal to be discovered about the feedbacks and linkages that allow these interconnected parts to function as a whole and, in turn, how they will respond to human modification.

Human Disturbance

Life influences the amount of CO_2 in the atmosphere through photosynthesis, respiration, and oceanic absorption. As ecosystems are altered, the balance of these processes will be altered. Human activities are upsetting this balance and increasing CO_2 in the atmosphere through the burning of fossil fuels and clearing of forests.A significant increase in CO_2 could have dramatic consequences.

Mathematical models of the climate suggest that when CO_2 doubles (sometime in the middle of the next century should population, economic and technology trends continue as typically projected), the world will warm up somewhere between 1 and 5 degrees Celsius by 2100 A.D. unless other factors counteract or amplify the CO_2-induced change Even the lower end of that range is a projected warming at the rate of one degree per hundred years, a factor of ten faster than the one degree per thousand that has been the typical average rate of natural sustained global temperature change from the end of the ice ages to warmer interglacial times. Should the higher end of the 1 to 5 degree Celsius range occur, then we could see rates of climate change some fifty times faster than sustained, natural average conditions. Climate largely determines the types of ecosystems that occupy an area. Global climate change at such a rapid rate would force many species to shift their ranges in an attempt to keep up with changing climatic conditions, as occurred during the ice age-interglacial transition ten to fifteen thousand years ago.

Migrations of species such as slow growing trees with large seeds would have to occur much faster than they did in the past to keep up with rapidly shifting climates. Other species could move more easily, raising the likelihood that communities of species could be disassembled. Estimating the rates of global warming in the next century, however, are very controversial because of the uncertainties involved with multiple interacting feedback mechanisms Humanity can control climate in ways other than changing greenhouse gas concentrations. Consider the amount of moisture released to the atmosphere through transpiration in the tropical rainforests. The dense vegetation in areas such as the Amazon basin typically recycles the precipitation that falls on it many times over, helping to form heavy cloud cover in the region. The clouds, in turn, reflect sunlight and produce more rain, directly influencing regional climate as well as indirectly perturbing global climate through altering large-scale circulation patterns over the tropics. As humanity deforests regions like the Amazon, not only is CO_2 released into the atmosphere, but changes in the hydrologic cycle will almost certainly affect regional climate and possibly even global climatic patterns. In deforested areas of northeastern Brazil, the cutting of the tropical forests has led to desertification, changing both surface reflectivity and the rate of transpiration. This change in ecosystem character can lead to a destabilizing positive feedback, which may cause an even further reduction in precipitation.

In a recent study on the possible climatic impacts of tropical deforestation, researchers suggest that conversion of forest into crop land or pastures would cause significant changes in the local microclimate Expected changes include reduction in soil moisture, larger diurnal fluctuation of surface temperature and humidity deficit, and increased surface run-off during the rainy season and decreased run-off during the dry season. Results from general circulation model simulations of large-scale deforestation and conversion to grassy vegetation in the Amazon basin indicate an increase in surface temperature, decrease in evapotranspiration, and significant reduction in precipitation.

Depending on the scale of the disturbed areas, local climate changes can lead to regional climate changes which, in turn, may cause alterations in the global climate through atmospheric connections between tropical circulation and large-scale circulation patterns outside of the tropics. The effect on the ecological systems through changes in the hydrological cycle, an increase in the dry season, and the disruption of plant-animal interactions may make it difficult for the rainforests to re-establish themselves if they are destroyed. Climate change aside, the implications of this scenario for the conservation of biodiversity are serious. The provision of fresh water and regulation of its flows through precipitation, evapouration, transpiration, and run off is mediated by all ecosystems. Forests and other vegetation types are critical components of this ecosystem service providing free flood and drought relief among other things. The loss of these services, through landuse change, can exacerbate disasters like spring floods in the Midwest and Southeast resulting from large expanses of land cleared for agriculture as well as the drainage of wetlands and swamps which otherwise might have acted as reservoirs for holding excess water or filtering toxic wastes.

Valuing Carbon

There is already a historic background on the evaluation of carbon that includes climate change policies such as the introduction of carbon taxes that reduce greenhouse gas emissions through increasing prices of carbon-based fuels proportional to the amount of carbon they emit Another mechanism for reducing greenhouse gas emissions is through an international tradable emissions permit system intended to limit emissions of certain pollutants. These policies and others constitute ways of balancing the economic costs of emissions with some assumed benefit of averting the loss of ecosystem services (called "climate damage").

Any comprehensive attempt to evaluate the societal value of climate change should include such things as loss of species diversity, loss of coastline from increasing sea level, environmentally displaced persons, and agricultural losses. first estimated the climate damage at 1% reduction in GNP based on market sector losses for a central estimate of climate change. This was criticized as too narrow a view of climate as a type of public good since it reflected neither non-market values (*e.g.* species loss) nor climate "surprise" scenarios conducted a survey of conventional economists, environmental economists, atmospheric scientists, and ecologists.

Their estimates of loss of gross world product (GWP) resulting from a 3 degree Celsius warming by 2090 varied between a loss of 0 and 21% of GNP with a mean of 1.9%). For a 6 degree Celsius warming scenario, the respondents predicted a loss of the world economy ranging from 0.8 to 62% with a mean estimate of 5.5%. A striking difference was noted between respondents from different academic disciplines, with natural scientists' estimates of economic impact 20 to 30 times higher than conventional economists'. Even a two per cent loss of GWP in 1995, however, represents climate damage of hundreds of billions of dollars annually.

While it is impossible to estimate credibly a numerical value on all of the ecosystem services provided through the maintenance of the carbon cycle at present state, it may be useful to look at land use change and loss of biomass, mostly through deforestation, as a source of atmospheric CO_2. In a very simplistic and preliminary evaluation, we can use the rates of net deforestation to calculate a value for carbon. Global loss of above-ground biomass from deforestation in the tropics is approximately 1-3 gigatons/year over the past 10 years. This amounts to between 2-5 gigatons of carbon in carbon dioxide released into the atmosphere each year from deforestation and forest degradation (this does not include the 6 gigaton carbon emissions from the burning of fossil fuels). Much of the carbon from biosphere emissions is taken up immediately by vegetation, however, leaving approximately 1-2.5 gigatons of net carbon added to the atmosphere each year. We can apply the concept of carbon taxation for emissions to an ecosystem service valuation of retaining the carbon in the forests.

Using a range of carbon taxes from typical macroeconomic models between $1 per ton and $100 per ton of carbon, the net value of carbon lost each year amounts to between $1 and $250 billion/year. However, use of optimizing economic models to estimate climate damage is highly unsatisfying since these studies use very limited and often ad hoc assumptions that both over and underestimate the likely damages to various market and non-market sectors.

Methods of Valuation

The need for alternative methods of evaluation of these climate-related ecosystem services is quite clear when examining preliminary public opinion responses of global warming. In a controversial method called contingent valuation respondents are surveyed to determine how much they would be willing to pay to prevent a given global climate change scenario from happening or accept if so much change were to be allowed. The difficulties with this type of valuing of environmental goods and processes are immense, especially since much of the evaluation is subjective. Public opinion depends, in part, upon people's exposure to the issues and the level of education and information on these issues they have received.

In a Southern California study, the contingent valuation technique was applied to the determine the influence of potential changes in temperature and precipitation resulting from global warming on respondents willingness to pay. Factorial survey methods were used to present a variety of hypothetical climate scenarios to a sample of 600 Southern California residents. Respondents were provided with a baseline microclimate for the region before future climate scenarios were evaluated. For residents living in coastal communities, the baseline climate over the past 10 years was described as having a summer average high temperature of 75 degrees Fahrenheit, with daily highs ranging between 70 and 80 degrees, and an average of 13 inches per year of rain. One possible future scenario over the next 10 years included a summer average high temperature of 100 degrees Fahrenheit, with daily highs ranging from 80 to 120 degrees (the latter typical of Death Valley, California), and an average of 20 inches per year of rain.

With these and other scenarios, predicted probabilities were determined from the respondents willingness to pay for the abatement of different mean high temperatures. In this scenario, respondents were willing to pay an average $140 to offset a mean high temperature of 100 degrees, while a mean high temperature of 80 degrees was worth approximately $100 This represents a 40% increment in willingness to pay for a 20 degree rise in temperature and other scenario characteristics. residents reached a plateau in their willingness to pay to prevent 120 degree Fahrenheit mean temperatures as compared to 110 degree Fahrenheit mean temperatures.

However, the actual damages to the L.A. basin residents of mean high temperatures of 110 or more degrees Fahrenheit (which would imply occasional extreme heat waves similar in temperature to Death Valley mean highs) would be orders of magnitude more costly, we believe, than a 100 degree Fahrenheit mean high temperature, as such extreme heat would decimate most existing vegetation and threaten the lives of tens of thousands of elderly and other persons vulnerable to heat stroke. For just such reasons, Berk and Shulman strongly caution against taking the dollar values from the survey literally or using them in cost-benefit analyses as they confound several source of value including stewardship and altruism. In addition, some of the climate increases were well above the range of current scientific estimates of greenhouse warming.

The survey was not done in conjunction with atmospheric scientists and climatologists who could provide more realistic climate scenarios or ecologists, public health officials, or others who could help the respondents realise what such warming might mean for trees, birds, or people. Contingent valuation of the hypothetical good is possible when people believe the survey scenario. We present this type of evaluation study to highlight how difficult it is to find acceptable methods to place values on the climatic components of ecosystem services. In this survey case, the background of the respondents as well as their (limited) prior knowledge of the impacts of greenhouse warming played a large role in the survey outcomes. At the same time, however, contingent valuation points out that people are willing to pay to preserve ecosystem services as well as the tremendous need for education to help citizens more realistically value climate and other environmental

The ongoing disturbances of the atmosphere that affect the biogeochemical and physical processes that determine the climate may influence human and natural systems in profound ways. We have attempted to outline a few of the major ecosystem services that are associated with climate and the atmosphere, as well as introduce the challenging task of quantifying, and ultimately monetizing, these services. Current monetized estimates of climate damage by the middle of the 21st century from typical climate change scenarios range from slight economic benefit to a trillion or more dollars lost annually, with most macroeconomic assessments assuming a one to two per cent annual loss to GWP from climate change. Moreover, the interacting processes and biogeochemical cycles occurring across a wide spectrum of scales lead to synergistic effects that are not usually considered and sometimes not even known when we attempt to disaggregate and value ecosystem services.

The deforestation of the Amazon basin is one example of interacting scales where land use change affecting local and regional climate may also produce a net global residual. Even if the mosaic of regional effects average themselves out globally, there could be residual effects arising from heterogeneous forcing of the climate in areas outside of the tropics (*i.e.* regional high concentrations of sulfate aerosols or tropospheric ozone).

Ecosystems both mediate and respond to the climate system through a variety of physical, biological, and chemical feedback cycles. The uncertainty of resulting synergisms and potential global effects, as exemplified in the Amazon basin, points to the important challenge of defining and understanding the processes that link species and ecosystems with climate. With increasing knowledge, we can better anticipate ecological responses under changing climate scenarios. Meanwhile, humanity continues to perform this potentially trillion dollar unnatural experiment on "Labouratory Earth".

AIR, CLIMATE AND WEATHER

The Earth's atmosphere is a blanket of gases approx-imately 350km (218 miles) thick. It is a large and complex system that interacts with the Sun, the land, and the oceans in order to produce both the Earth's weather and climate.

The Earth's Atmosphere has four distinct layers:

1. Thermosphere
2. Mesosphere
3. Stratosphere
4. Troposphere

Thermosphere

The layer of atmosphere most distant from the Earth is the thermosphere, which begins approximately 80km in altitude. It is also the hottest layer; "thermo" being Greek for heat. The temperatures in the thermosphere increase with altitude due to the absorption of intense solar radiation by the limited amount of remaining molecular oxygen. The source of this heat is through bombardment of solar particles carried on the solar wind that do not reach deeper into the atmosphere.

Mesosphere

The mesosphere extends from 50 to 80km in altitude with very sparse atmosphere, accounting for only about 0.1 per cent of the mass of the atmosphere as a whole. Temperatures decline within the mesosphere as altitude rises, containing the coldest temperatures within the Earth's atmosphere. At its upper boundary, the mesopause, average temperatures are near -110°C in the summer and -60°C in the winter.

Stratosphere

The stratosphere extends from approximately 10-12 km to around 50 km above the Earth's surface. The air temperature remains relatively constant up to an altitude of 25 km, then increases gradually having a stabilizing effect on atmospheric conditions. The stratosphere contains nearly 90 per cent of the atmospheric ozone, which plays a major role in regulating temperatures as solar energy is converted to kinetic energy when the ozone molecules absorb ultraviolet radiation, resulting in the heating of the stratosphere.

Troposphere

The troposphere is the layer closest to the Earth's surface, containing more than 80 per cent of total atmospheric mass composed of 78 per cent nitrogen, 21 per cent oxygen, other trace gases, water droplets, dust, and other particles. The troposphere is where most weather occurs; the circulation of air in intensive vertical movements results in the formation of clouds, while horizontal movements results in wind. Both temperature and water vapour content decrease rapidly as altitude increases within the troposphere. Nearly 99 per cent of atmospheric water vapour is contained within this layer, which plays a major role in regulating air temperature as it absorbs solar energy and thermal radiation from the planet's surface. Scientists believe that the Earth's climate is changing and is, in fact, heating up. However, there are considerable differences among the views with regard to the rate of change, the impact on our environment, and what can or should be done about it. Current environmental concerns are increa-singly focusing on the particles and gases released into the atmosphere as a result of human activity. Frequently, unusual weather events, such as Hurricane Katrina in 2005, are being linked to global warming. But "weather" and "climate" are two very different things, and it is important to understand the difference, as well as the relation, between the two.

In the troposphere, air rises as it is heated by the sun, falls earthward as it cools, and intermixes with evapourated water from the planet's bodies of water to form clouds and precipitation. The uneven heating of the Earth's surface by the sun, along with Earth's rotation, creates rising (convection), falling (advection), and horizontal air movements (winds). The result of these processes occurring in the form of rain, snow, heat or freezing cold, at a particular place and time, is called weather. The longer term trends in patterns of temperature, rainfall, and other weather indicators over time, usually in blocks of 30 years that can affect the entire Earth, is called climate.

Land and Land Use

Land is the solid part of the Earth's surface, also known as the lithosphere. The surface of the Earth is shaped by a combination of physical processes, including earthquakes and volcanoes, shifts of rocks and sediments, and flows of river and ice. Human activities also shape the land in many ways, including expanding agriculture and population, excavating mineral and forest resources, and changing the flow of rivers with dams and channels. Land cover is the physical and biological material found on the land surface – which can be either vegetation or man-made structures. Land cover differs from land use, although the way land is used is often either a cause or an effect of its cover type. Land use describes the various ways in which human beings make use of and manage the land and its resources. The way land is used and managed can alter land cover, and can also have an impact on surrounding habitats, the flow of surface waters, and on biogeochemical cycles.

Land cover and land use issues, therefore, can play a significant role in environmental change, including climate change. When land cover is altered, it can result in the destruction of natural carbon sinks that absorb atmospheric carbon dioxide, a heat-trapping gas that contributes to climate change. Deforestation, is a major contributor to observed increases of carbon dioxide in the atmosphere. There are often competing interests and tradeoffs related to land use that must be taken into consideration. Land use planners and decision makers often struggle with how to meet human needs for food and the utilization of our natural resources while also minimizing impacts and preserving important habitats. These choices are typically made by those who own or control the land; although the choices can also be limited by the physical and biological characteristics of the land, including the climate, soils, and geography. Land use choices are also limited by institutional factors and, in the United States, can be governed by a mix of local, county, state, and federal laws. Although most of the land. is privately owned, the federal government is the single largest owner of land. Analysis of land use and land cover has been greatly facilitated in the last few decades with the availability of new technologies, including remote sensing. The development of geographic information systems (GIS) has made it possible to combine physical characteristics of the land with information about population, economic data, and other information for further analysis

Water

"Water, water, everywhere, nor any drop to drink." More than 70 per cent of the Earth's surface is covered by water. When the Apollo space missions beamed back pictures of the Earth, public attention was caught by the image of a "Blue Planet," because of the vast expanses of ocean covering the globe. Yet despite this seemingly inexhaustible supply (approximately 370 billion billion gallons) of water on earth, access to clean drinking water remains a major challenge for a large percentage of the world's population. Almost 97 per cent of the Earth's water is in the oceans and is too salty to drink. Less than 1 per cent of the water on the earth is fresh water (the remaining two to three per cent is frozen fresh water contained in glaciers and ice caps). Water is indispensable for life, not only for drinking water, but also for raising crops and animals for food. Fishing provides a major source of protein for much of the world's population. Throughout history, cities and villages have grown up near sources of water, for drinking, sustenance, and transportation. In this century, population growth has increased in areas where fresh water is relatively scarce. Water in these areas is a valuable commodity and, in some cases, great technological feats have been undertaken to provide a reliable source of water.

In some areas, water quality is a serious concern and waterborne diseases remain a significant source of mortality in many developing countries. The United States has made considerable progress over the last few decades in reducing direct discharges of pollutants into lakes, rivers, and streams. However, indirect sources of pollutants, including both nutrients and toxic substances that are carried by flowing surface waters into larger bodies of water, have proven more difficult to control.

Ecosystems

An ecosystem is generally defined as a community of organisms living in a particular environment and the physical elements in that environment with which they interact. But where does one particular ecosystem end and another begin? An ecosystem can be as small as a field or as large as the ocean, depending on the scale that the researcher is examining. The borders of an ecosystem may be clear, such as a pond; other borders may be less easy to define, such as grassland that gradually changes into brush.

Just as there is an immense diversity of individual species on the planet, so is there a rich diversity of ecosystems, from the icy arctic zones to tropical forests lush with plants and animals. Even the depths of the oceans, once thought to be barren, are now known to be teeming with living microorganisms and other life. There is much that remains to be discovered. Biologists do not know with any certainty how many species there are or even why some areas, such as the tropics, are richer in biological diversity than others.

It is known that human activities, mostly unintended, threaten biodiversity by altering habitats and introducing non-native species. There is general agreement on the importance of protecting biological diversity, especially since humans depend on the services provided by living organisms and ecosystems. There is less agreement on the best approaches to conservation and how to balance preservation of habitats with meeting human needs.

Energy

The development of modern civilization has been dependent on both the availability and the advancement of energy. We have witnessed a progression from animal and steam power, to the internal combustion engine and electricity generation and to the harnessing of alternative sources of energy. Because of our reliance on energy sources, it is also important to understand the impacts of energy use on the environment. All aspects of energy the way it is produced, distributed, and consumed can affect local, regional, and global environments through land use and degradation, air pollution, the acidification of water and soils, and through global climate change via greenhouse gas emissions.

The majority of our current energy stems from fossil fuels such as coal, oil, and natural gas. Coal dominates in our production of electricity, while oil is the world's primary transportation fuel. Natural gas use, which is most commonly as heating fuel, is currently growing faster than other fossil fuels. While cleaner and less carbon intensive than both coal and oil, its use still emits significant amounts of carbon dioxide.

Over the foreseeable future, it is very likely that fossil fuels will remain our largest source of energy. However, fossil fuels are finite resources and there is concern not only about both domestic supply and reliance on foreign supplies, but also with the increasing cost of these fuels. The burning of fossil fuels is also the largest source of carbon dioxide emissions, which contributes to the greenhouse effect. As these fuels continue to decrease in supply and impact the environment, there will be an increasing need to utilize alternative fuel sources.

Nuclear energy provides nearly a fifth of the world's electricity and does not produce harmful by-products, only non-radioactive water vapour. However, controversy remains over the use of nuclear energy due to concerns over safe storage and disposal of radioactive waste, and the potential for accidents and radiation contamination and exposure.

Hydrogen is the most abundant element in the universe and could be an important factor in our energy future. Hydrogen fuel cells can produce power without emitting any pollutants; their only byproducts are water and heat. Hydrogen can both carry and store energy and can be used in a wide variety of applications, including portable devices that use batteries, transportation vehicles, and a number of stationary power sources. The majority of hydrogen is currently produced by processing fossil fuels (which emits pollutants through the initial process), but it can also be manufactured from renewable energy sources and feedstocks through the electrolysis of water. With the increased cost and environmental impacts related to fossil fuel use, along with the controversy over nuclear power, research and development in the area of renewable sources of energy continues to flourish. Alternative sources – such as wind, solar, geothermal, and water – have been used in one form or another for many centuries, but will require additional advancement before they can become truly cost-competitive with conventional energy sources.

Wind is an abundant, inexhaustible, and clean source of energy; and, advancements in the design of wind turbines has lowered the cost of wind-generated electricity. However, wind variability, noise pollution caused by the turbines, and the potential to harm birds continue to be areas of concern.

Solar technology obtains power from sunlight. There are three types of solar technology: solar electric, solar hot water, and concentrating solar power. The cost of solar power has declined over the past 30 years, and will continue to decline with increased technological efficiency of technology and manufacturing improvements. However, it is also subject to changes in weather (cloud cover) and pollution, which can block sunlight and decrease overall productivity. Geothermal energy taps into heat underneath the Earth's crust to boil water that is then used to drive electric turbines to heat buildings, homes, or in other non-electrical purposes. Over the past two decades, geothermal energy costs have decreased nearly 50%. However, while abundant, in only a few locations is it close enough to the Earth's surface to provide for large scale power.

Water–or hydroelectric–power has been harnessed for thousands of years, beginning with the water wheel. For more than a century, the technology for using falling water to create hydroelectricity has evolved to its current use of turbines and generators to convert the energy into electricity, which is then fed into the electric grid. Although a widely accessible energy source in many countries–concerns over land use, impacts on fish populations, and water flow and quality remain. One effective way of reducing environmental impacts from the current production of energy is to increase energy conservation. By using energy more efficiently, both costs and emissions are lowered; therefore, a significant incentive exists to improve efficiency in manufacturing processes, the construction of new buildings, the production of new appliances, and for many other energy uses.

Several developed countries are already taking action to increase energy efficiency and, on a larger scale, new measures are being explored to assist developing countries with energy conservation as they continue to increase their energy consumption. Under the Kyoto Protocol, developed countries can provide technological assistance through "clean technology" projects with developing nations and can even put the emission reductions towards compliance in their own countries. As human consumption of energy increases and we come closer to using up our supply of fossil fuels, further research and development will be necessary to produce alternative and/or renewable sources of energy that are readily available, affordable, and less harmful to the environment. While our dependence on energy is not likely to decrease, there will continue to be an increase of new innovations in energy technologies with a larger focus on energy conservation and efficiency.

Energy Information Administration (EIA): The EIA also provides information on the various types of energy, including renewable, nonrenewable, and secondary sources. Congressional Research Service: Energy Issues Policy papers on energy issues can be found on this site, including articles and several reports related to energy policy and environmental protection.

Food

Human beings began domesticating plants and animals as long as 10,000 years ago. Success in this endeavor led to drastic changes in how and where human beings lived. It changed how humans interacted with each other and how they interacted with the earth. Agriculture led humans from a nomadic existence to one based in permanent and semi-permanent settlements. Human civilization has its roots in the early domestication of plants and animals.

The unprecedented growth in global population that occurred in the twentieth century was made possible by remarkable advances in agriculture and technology. The Green Revolution brought high-yield crops and advanced growing techniques to developing countries and improved nutrition and health for most parts of the world. Continuing population growth will require more acres of land to plant and more water for irrigation, increasing pressure on habitats and natural resources. Advanced agricultural methods have made it possible to grow more food on fewer acres of land, which permits much land to be returned to forest and other uses. However, there are costs associated with high-yield methods, including their heavy reliance on pesticides and fertilizers.

Run-off of fertilizers from agricultural lands has affected the water quality of some lakes, rivers, and bays. Erosion is, and has been since crops were first planted, a serious environmental problem in the absence of preventative measures. Irrigation, required in many parts of the world, can be a significant drain on water supplies in arid areas and, if improperly managed, can lead to buildup of salt deposits, which may severely degrade soil quality.

One of the most critical challenges of the next few decades is to find ways to increase production of food while minimizing environmental degradation. Considerable research in this and other countries has led to advanced methods to preserve soil and prevent erosion, including conservation tilling and computer-controlled application of pesticides, fertilizers, and water. Pesticide use is controversial. Other methods, especially new types of genetic modification, or "genetic engineering" are more controversial.

To reduce reliance on pesticides, methods such as biocontrol (using other species to control pests) and biotechnology to create pest-resistant crops are being tested. Biocontrol using wild-type (naturally occurring) species is usually accepted, especially by organic producers. Naturally occurring mutants are accepted as well. More controversial is the use of genetically modified biocontrol agents changed by DNA deletion or insertion of new traits. Concerns about the potential impacts of these methods have been raised.

Environment and Society

Throughout history humans have both affected, and been affected by, the natural world. While a good deal has been lost due to human actions, much of what is valued about the environment has been preserved and protected through human action. While many uncertainties remain, there is a realization that environmental problems are becoming more and more complex, especially as issues arise on a more global level, such as that of atmospheric pollution or global warming.

Interactions between human society and the environment are constantly changing. The environment, while highly valued by most, is used and altered by a wide variety of people with many different interests and values. Difficulties remain on how best to ensure the protection of our environment and natural resources. There will always be tradeoffs and, many times, unanticipated or unintended consequences. However, a well-managed environment can provide goods and services that are both essential for our well being as well as for continued economic prosperity.

4

Seafood Preservation Techniques

Ancient methods of preserving fish included drying, salting, pickling and smoking. All of these techniques are still used today but the more modern techniques of freezing and canning have taken on a large importance. Fish curing includes and of curing fish by drying, salting, smoking, and pickling, or by combinations of these processes have been employed since ancient times. On sailing vessels fish were usually salted down immediately to prevent spoilage; the swifter boats of today commonly bring in unsalted fish. Modern freezing and canning methods have largely supplanted older methods of preservation. Fish to be cured are usually first cleaned, scaled, and eviscerated.

Fish are salted by packing them between layers of salt or by immersion in brine. The fish most extensively salted are cod, herring, mackerel, and haddock. Smoking preserves fish by drying, by deposition of creosote ingredients, and, when the fish are near the source of heat, by heat penetration. Herring and haddock (finnan haddie) are commonly smoked. Kippers are split herring, and bloaters are whole herring, salted and smoked.Sardines, pilchards, and anchovies are small fish of the herring family, often salted and smoked and then preserved in oil. Fish are dried under controlled conditions of temperature, humidity, and air velocity. Since the dried product is relatively unappetising and rehydrating slow, other preservation methods are common. Many different techniques have been used to preserve fish quality and to increase their shelf life. They are designed to inhibit or reduce the metabolic changes that lead to fish spoilage by controlling specific parameters of the fish and/or its environment. These techniques can be classified as follows: These encompass a wide array of technologies used to decrease the fish temperature to levels where metabolic activities - catalysed by autolytic or microbial enzymes - are reduced or completely stopped. This is possible by refrigeration or freezing where the fish temperature is reduced, respectively, to approximately 0 °C or < - 18°C.

Fish refrigeration can use cool air circulating around the fish (mechanical refrigeration) or icing.Fish icing and boxing on-board fishing vessels is not always possible in the case of small pelagics that are caught in large quantities. These are chilled using refrigerated seawater (RSW) or chilled seawater

(CSW). Chilled or frozen fish products require additional cooling in cold store to avoid an increase in temperature. The design (size, insulation, palletisation) and management of cold stores are key for fish quality and energy saving. A major environmental issue relates to the development of alternative refrigerants to replace the chlorofluorocarbons (CFCs) which are damaging to ozone layers.

Techniques Based on the Control of Water Activity

Water activity (a_w) is a parameter that measures the availability of water in fish flesh. It is expressed as the ratio of water vapour pressure in fish/vapour pressure of pure water at the same temperature and pressure. A_w varies from 0 to 1.

Fig. Chorkor Oven for Smoking Fish.

Water is necessary for microbial and enzymatic reactions and several preservation techniques have been developed to tie up this water (or remove it) and thus reduce the aw. These include drying, salting, smoking, freeze-drying, the use of water binding humectants and a combination of these. Some of these techniques, such as drying, salting and hot smoking, have been used for thousands of years. They can be implemented very simply, *e.g.* by salting, solar drying, or using fully automated equipment with temperature control, relative humidity, etc.

Techniques Based on the Physical Control of Microbial Fish

These physical methods use heat (cooking, blanching, pasteurising, sterilising), ionising irradiation (for pasteurisation or sterilisation) or microwave heating. Cooking or pasteurising are processes that do not allow complete inactivation of microorganisms and thus often need to be combined with refrigeration to preserve fish products and increase their shelf life. This is not the case of sterilised products and which are stable at ambient temperatures (< 40°C). These require packaging in metal cans or retortable pouches before the heat treatment, thus the term "canning".

Techniques Based on the Chemical Control of Microbial Activity and Loads

These techniques are designed to add anti-microbial agents or decrease the fish muscle pH to levels that are inhibitory to microbial growth and proliferation. Most bacteria stop multiplying at pH < 4.5. The decrease of pH is obtained by fermentation, marinades or by adding acids (acetic, citric, lactic, etc.) to fish products.

Fig. Fog/ Drying Fish.

In addition to the decrease in fish pH, fish fermenting lactic bacteria also produce anti-microbial compounds such as nisin, which improve preservation. This technique is often referred to as bio-preservation. Other preservatives include nitrites, sulphites, sorbates, benzoates or natural ones such as essential oils.

Techniques Based on the Control of the Oxydo-Reduction Potential

Some spoilage bacteria and lipid oxidation require oxygen. Reducing the oxygen around fish will increase its shelf life. This is possible by vacuum packaging or by controlling or modifying the atmosphere around the fish. Specific combinations of CO_2, O_2 and N_2 characterise controlled (CA) or modified atmosphere (MA). Vacuum packaging, CA and MA storage are often combined with refrigeration for fish preservation.

Combination of Several Preservation Techniques

Two or more of the above-described techniques can be combined to improve preservation efficiency while reducing undesirable effects such as the denaturation of nutrients by severe heat treatments. Combinations already in use include pasteurisation-refrigeration, CA (or MA)-refrigeration, salting-drying, salting-smoking, drying-smoking and salting-marinating. Other process combinations are currently being developed along the "multiple hurdle theory".

SMOKING SEAFOOD: AN ANCIENT METHOD OF PRESERVATION

Smoking food is a cooking method that has existed since man discovered fire and found his smoke-filled cave also served to preserve fish and meat.

Smoking was also used to prepare food for storage over the winter when fresh meat and seafood wasn't readily available. In the Pacific Northwest Native Americans smoked fish using alder sticks stuck into the ground alongside a fire or on a drying rack over a smoldering fire.

Commercial smoking of seafood began on a large scale in the 1800's with scores of smokehouses opening in Brooklyn, New York bringing salmon in from the west coast and later using North Atlantic salmon. Unfortunately heavy commercial fishing of salmon almost drove the fish into extinction. Today most salmon comes from the Pacific Northwest or the North Atlantic while some is imported from Chile and Norway.

Cold-smoked salmon - sliced very thinly and eaten raw - is an elegant dish. Lox (smoked salmon that is also cold-smoked) is more heavily cured and is often eaten on a bagel with cream cheese, red onion and capers. Kippers are cured and smoked herrings, very popular in Britain for breakfast and lunch. Residents of Scotland's village of Findon are credited with creating finnan haddie (smoked haddock that is either poached or broiled after smoking and served hot) that was originally smoked over peat. Halibut, sturgeon and fish roe are usually cold-smoked while eel, trout, bluefish and mackerel are hot-smoked and can be eaten without further cooking.

There are two ways of curing seafood: using a brine solution or a dry cure. Most store bought smoked salmon is brined. The smoking process also varies (as does the smoking of meats) and two methods are used: hot-smoking or cold-smoking. The main difference is the length of the smoking and the temperature used for smoking. Cold smoked seafood tends to be slightly smoky and oilier while the hot-smoked version is much drier with a stronger smoke flavour. Temperatures for cold smoking are typically between 68 to 86 degrees F. At this temperature range foods take on a smoked flavour but remain relatively moist. Cold smoking does not cook foods and frequently takes days of smoking. Hot smoking exposes food to smoke and heat in a controlled environment. Although foods that have been hot smoked are often reheated or cooked, they are safe to eat without further cooking. Hot smoking is done between 126 to 176 degrees F. while barbecue, also called smoke roasting or pit roasting, is traditionally done at 225 to 250 degrees F. Most any kind of wood can be used for smoking as long as it's non-resinous. Typical woods are fruit woods, alder, oak, maple, and hickory. Salmon is usually smoked with alder. Although any fish can be smoked, fish with higher oil content (salmon, mackerel, sturgeon, bluefish, trout, eel, herring or smelts) are better suited as they absorb more smoke and remain moist during cooking.

PREPARING FISH

Use only fresh fish that has been kept clean and cold. Salmon are split with the backbone removed or filleted; bottom fish filleted; herring and smelts are headed and gutted. (Herring are also traditionally split for kippers.) Rinse the fish with running cold water to remove all traces of blood.

Basic Fish Brine

- 4 cups cold water
- 1/4 cup (2.5 ounces or 70 grams) kosher salt
- 1/4 cup dark brown sugar, packed
- 2 bay leaves
- 1 stalk celery, sliced
- 1/2 cup chopped yellow onion
- 2 cloves garlic, smashed

Bring two cups of water to a simmer over medium heat. Add the remaining ingredients and stir until the salt and sugar dissolve. Remove from the heat and add the remaining cold water. Cool the brine to 40 degrees F. (Use one gallon of brine for every four pounds of fish.) If you don't want to make your own brining solution or want a variety of flavoured brines, Hi Mountain offers several brine mixes including a trout brine, gourmet fish brine and Alaskan salmon brine. Place the fish in a non-reactive container and add the brine. Cover and refrigerate. Brine 1/2-inch thick fillets for two hours, one-inch thick pieces about six hours, and 11/2-inch thick pieces overnight. Brining times can be adjusted to give the fish a lighter or heavier cure. Brining pickles the fish so the longer it brines the saltier it will get. Another option is to dry cure the fish instead of using a brining solution.

Basic Dry Cure

- 1/2 cup (5 ounces or 140 grams) coarse kosher salt
- 1/2 cup dark brown sugar, packed
- 3/4 teaspoon cracked black pepper
- 1 1/2 teaspoons dried lemon zest
- 1 teaspoon granulated garlic
- 1/2 teaspoon dried ginger

Combine all the ingredients in a mixing bowl and mix well to combine. (This cure will keep in a closed container for six months.) Liberally apply the cure to both sides of the fish. Place in a non-reactive container and cover. Refrigerate for six to eight hours. After brining or curing, rinse the fish with cold running water. Place the fish skin side down on a cooling rack over a half sheet pan and refrigerate overnight allowing a pellicle to form on the surface. The pellicle, a sticky lacquer-like layer, will seal the surface and prevents loss of natural juices during smoking.

Prepare the smoker for 150 degrees F. smoking and smoke the fish using the wood of your choice to an internal temperature of 140 degrees F. (Generally 1/2-inch pieces are smoked for an hour; one-inch pieces for two hours and 1 1/ 2-inch pieces for three hours.) Use an instant-read thermometer such as the Thermapen to assure the fish is properly cooked. Once the fish is smoked, wrap the cooled fish with plastic wrap. It will keep in the refrigerator for a week or frozen for up to six months.

SEAFOOD PRESERVATION PROCESS

The preservation of meat products is accomplished utilising a combination of smoke, ozone and freezing preservation techniques. Particularly, fish products are sized into portions that are first treated with smoke, followed by treatment with ozone and then optionally frozen.

The preservation system extends the shelf life of the fish products and permits the fish to maintain its freshness and freedom from bacterial decomposition for a longer period of time following catch. The preservation process further maintains the characteristics of day caught fish, such as taste, texture and colour, making the refreshed fish products produced by the present system more appealing to consumers.

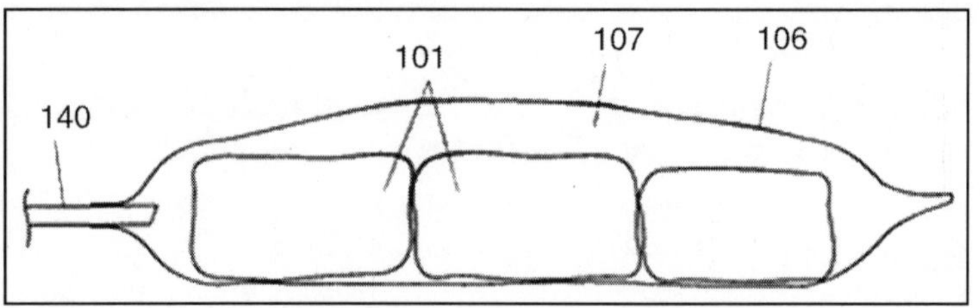

Fig. Representative Image.

CROSS REFERENCE TO RELATED APPLICATION

This application claims priority to U.S., provisional application entitled, "Seafood Preservation Process," having Ser. No. 60/241,920, filed Oct. 20, 2000 now abandoned, which is entirely incorporated herein by reference. Claims:

Therefore, having thus described the invention, at least the following is claimed:

- A process for treating raw fish products for increasing the shelf life of the fish products and preserving the colour and freshness of the fish products after the fish products have been frozen and thawed, comprising the steps of: subjecting the raw fish products to smoke to apply smoke to the fish products for a period sufficient to preserve and substantially maintain the colour of the flesh and blood line of the fish products; a removing the smoke from about the smoked raw fish products; subjecting the smoked raw fish products to ozone to apply ozone to the smoked fish products for a period sufficient for the ozone to substantially remove smoke odour from the fish products and sufficient to reduce the existing bacteria on the fish products; and after the steps of applying smoke and ozone to the raw fish products freezing the raw fish products for storage and shipment, so that the combination of the smoking and ozone applications is sufficient such that when the smoked and ozone treated raw fish products are thawed, the colour of the fish products is substantially maintained and the shelf life of the thawed fish is extended compared to fish that have been similarly treat with smoke, frozen and thawed without the ozone application.
- The process of claim 1, further including the step of before the fish products are subjected to smoke and ozone cutting the fish products into portions that are of a size that the smoke applied to the fish products will permeate into and substantially through the fish products.
- The process of claim 1, wherein the step of subjecting the fish products to smoke comprises subjecting the fish products to smoke for a sufficient period of time such that the smoke permeates the fish products and serves as a preservative of the fish products.
- The process of claim 1, and further including the additional step of freezing the fish products before the fish products are subjected to smoke and ozone, and wherein the step of subjecting the fish products to smoke comprises simultaneously thawing the fish products and applying smoke to the fish products.
- The process of claim 1, wherein the step of subjecting the fish products to smoke comprises cleaning the smoke by removing a portion of the particulate in the smoke prior to applying the smoke to the fish products.

- The process of claim 1, wherein the step of subjecting the fish products to smoke comprises placing the fish products in a container, reducing the pressure in the container to a pressure less than atmospheric pressure, and introducing smoke into the container when the pressure in the container is less than atmospheric pressure.
- The process of claim 6, and further including the step of, before the step of subjecting the fish products to smoke, cleaning the smoke by removing a portion of the particulate in the smoke before the fish products are subjected to the smoke.
- The process of claim 1, wherein the step of subjecting the fish products to smoke comprises burning organic matter to form smoke, cleaning the smoke, chilling the smoke, and subjecting the fish products to the chilled and cleaned smoke at a pressure other than atmospheric pressure.
- The process of claim 1, wherein the step of subjecting the smoked fish products to ozone comprises subjecting the fish products to ozone through a carrier medium.
- The process of claim 9, wherein the ozone is applied to the flab products in a containment chamber.
- The process of claim 1, wherein each of the steps of subjecting the fish products to smoke and ozone comprises reducing the atmosphere about the fish products to a pressure less than atmospheric pressure and introducing the smoke and ozone to the fish products in the reduced pressure atmosphere.
- A process for treating raw fish for increasing the shelf life of the raw fish and preserving the colour and freshness of the raw fish after the raw fish have been treated, frozen and thawed, comprising:
  - Forming raw fish into fish products of a predetermined size
  - Placing the fish products into a chamber;
  - Withdrawing air from about the fish products in the chamber;
  - Generating smoke from an organic substance;
  - Removing the heavier particles from the smoke to create cleaned smoke;
  - Introducing the cleaned smoke into the chamber and applying the cleaned to smoke to the fish products in the chamber;
  - Containing the cleaned smoke within the chamber with the fish products for a predetermined period sufficient for the smoke to permeate into and substantially through the raw fish products, and to substantially maintain the colour of the fish products;
  - Withdrawing the cleaned smoke from the chamber while the fish products remain in the chamber;

- Introducing ozone into the chamber and applying the ozone to the smoked fish products in the chamber;
- Containing the ozone within the chamber with the fish products for a period sufficient to reduce the existing bacteria on the fish products and to substantially remove the aroma and taste of smoke from the fish products;
- Withdrawing the ozone from the chamber; and
- Freezing the smoked, ozone treated raw fish products, so that the combination of smoking and ozone treatments of the raw fish products is sufficient such that when fish products are thawed, the thawed fish products have substantially retained their colour from the time prior to having been smoked and the shelf life of the thawed fish products is extended compared to other similarly smoked, frozen and thawed fish without the ozone application.

- The process of claim 12, and further including repeating steps f and h before performing step i.
- The process of claim 13, and further including the step of repeating steps i, j, and k.
- The process of claim 12, wherein the step of withdrawing air from about the fish products in the chamber comprises forming a pressure less than atmospheric pressure in the chamber.
- The process of claim 12, wherein the step of containing the cleaned smoke within the chamber comprises placing the fish products in a vacuum bag.
- The process of claim 12, wherein the step of containing the cleaned smoke within the chamber comprises containing the cleaned smoke within the chamber and about the fish products for a sufficient time to apply enough cleaned smoke to the fish products so that the cleaned smoke on the fish products deters deterioration of the fish products.
- The process of claim 17, further including the step of monitoring the CO/CO_2 levels of the cleaned smoke residing within the chamber, and wherein the step of containing the cleaned smoke within the chamber with the fish products comprises containing the cleaned smoke in the chamber until an appreciable decline is noted in the CO/CO_2 levels.
- The process of claim 17, wherein the steps of introducing cleaned smoke into the chamber and containing the cleaned smoke in the chamber and about the fish products comprises removing the cleaned smoke from the chamber and introducing more cleaned smoke into the chamber and about the fish products until the colour characteristics of the fish products in the chamber have stabilised.

- The process of claim 12, wherein the step of introducing the ozone into the chamber comprises introducing the ozone into the chamber as a liquid solution containing ozone.
- The process of claim 12, wherein the step of introducing the ozone into the chamber comprises introducing the ozone into the chamber as a gas or gaseous mixture containing ozone.
- The process of claim 12, wherein the steps of introducing and containing the ozone in the chamber comprises containing the ozone within the chamber and about the fish products for a sufficient time to apply enough ozone to the fish products so that odour and taste of smoke in the fish products are no longer detectable.
- The process of claim 12, wherein the step of freezing the fish products comprises freezing the fish products after the steps of ozonation and removal of the fish products from the chamber.
- The process of claim 12, wherein the step of freezing the fish products comprises freezing the fish products after the steps of smoking, ozonation and removal of the fish products from the chamber.
- The process of claim 12, further including the additional step of freezing the fish products before the steps of placing the fish products into the chamber and prior to the steps of smoking and ozonating the fish products.
- A process of treating raw fish products for increasing the shelf life of the fish products and preserving the colour and freshness of the fish products at a time alter the fish products have been frozen and thawed, comprising the application of smoke to the raw fish products for a period sufficient to preserve and substantially maintain the colour of the raw fish products, followed by the application of ozone to the fish products for a period sufficient to substantially remove the aroma and taste of smoke from the fish products and to reduce the existing bacteria on the fish products, freezing of the raw, smoked and ozone treated fish products, maintaining the frozen fish products in a frozen state until use is desired, and thawing the fish products, so that the combination of smoking, ozone treatment and freezing of the raw fish products is sufficient such that when the fish products are thawed, the thawed fish products have substantially retained their appearance from the time prior to having been smoked and the shelf life of the thawed fish products is extended compared to other similarly smoked, frozen and thawed fish products without the ozone treatment.
- A process for preserving fish products as in claim 26, wherein the applications of smoke and ozone preserve the fish products by substantially preventing or inhibiting the growth of bacteria on the surface of and in the flesh of the fish products.

- A process for preserving fish products as in claim 27, further comprising the additional step of freezing the fish products before the application of smoke and ozone.
- A process for treating raw fish products for preserving the colour and freshness of the fish products and for increasing the shelf life of said fish products after the fish products have been frozen and thawed, comprising:
  - Placing a raw fish product in a vacuum bag;
  - Subjecting the fish product to smoke in the vacuum bag until the exposure of the fish product to the smoke substantially stabilises the colour characteristics of the fish product;
  - Removing the smoke from about the fish product;
  - Treating the raw fish product wit ozone for a period sufficient to substantially remove the aroma and taste of smoke from the fish products and sufficient to reduce the existing bacteria on the fish product; and
  - Freezing the raw, smoked and ozone treated fish product for storage and shipment, so that the combination of smoking and ozone treatment of the raw fish product is sufficient such tat when the fish product is frozen and later thawed the fish thawed fish product substantially retains its appearance from prior to having been smoked and the shelf life of the thawed fish product is increased compared to other fish that were similarly smoked, frozen and thawed without ozone treatment.
- The process of claim 29 and further comprising the step of monitoring the decline of CO/CO_2 levels in the smoke in the vacuum bag until the decline slows, at which point removing any remaining smoke from the vacuum bag.
- The process of claim 30 and further comprising the steps of continuing to subject the fish products to smoke and removing the smoke from around the fish products.
- The process of claim 29 wherein the fish products comprise laterally compressed fish products, the step of treating the fish products with ozone is performed by immersion of the fish products in an ozone dipping tank for a sufficient period of time to remove any detectable smoke odour or taste from the fish products.
- The process of claim 29, wherein the fish products comprise pelagic fish products, the step of treating the fish products with ozone is performed by immersion of the fish products in an ozone chamber for a sufficient period of time to remove any detectable smoke odour or taste from the fish products.
- The process of claim 29, wherein the fish products comprise laterally

compressed fish products, the fish products are frozen at a temperature below –35° C., stored at a temperature below –18° C. and maintained at a refrigeration temperature between approximately 2 and 5° C.

- The process of claim 29, wherein the fish products comprise pelagic fish products, the fish products are frozen at a temperature below –40° C., stored at a temperature below –18° C. and are thawed and maintained at a refrigeration temperature between approximately 2 and 5° C.
- A process of treating raw fish products for increasing the shelf life of the fish products and preserving the colour of the fish products after the fish products have been frozen and thawed, comprising the steps of: applying smoke to a raw fish product for a period sufficient to preserve the fish product and sufficient to substantially maintain the colour of the blood line of the fish product; after the smoke has been applied to the fish product, applying ozone to the smoked raw fish product for a period sufficient to substantially remove the aroma and flavour of smoke from the fish product and sufficient reduce the existing bacteria on the fish product; and freezing the raw, smoked and ozone treated fish product for storage and shipment, so that the combination of smoking, ozone treatment and freezing of the raw fish product is sufficient such that when the frozen fish product is thawed the fish product substantially retains its appearance from prior to having been smoked and the shelf life of the thawed fish product is increased compared to other fish that were similarly smoked, frozen and thawed without the application of ozone to the fish.
- A process of preserving the colour and freshness of previously smoked raw fish and increasing the shelf life of the smoked raw fish after the fish has been frozen and thawed comprising: treating previously smoked raw fish with ozone for a time sufficient to substantially remove the smoke aroma and flavour from the fish and sufficient to reduce the existing bacteria on the fish, and freezing the smoked, ozone treated raw fish, so that when the frozen fish is thawed the fish substantially retains its appearance from prior to having been smoked and the shelf life of the thawed fish is increased compared to other fish that were similarly smoked, frozen and thawed without the ozone treatment.

TECHNICAL FIELD

The present invention is generally related to the preservation of seafood and other food products for consumer consumption, and more particularly is related to a process for preserving fish by treating fish with smoke and ozone to retard degradation of the fish and maintain the fresh-like appearance of the fish. Optionally, the fish can then be frozen to further prolong its shelf life

Background of the Invention

The preservation of fish has been a major concern for fishermen and fish processors for centuries. Originally man salted and dried fish to preserve it. Since the advent of mechanical refrigeration, the fish have been preserved by freezing and refrigeration, thus permitting fishermen to make longer fishing trips, as well as transport the fish long distances over land or water.

The length of time over which fish maintains its freshness is commonly referred to as its shelf life. The shelf life of fish is determined by a number of factors, including the total number of each type of bacteria initially present, the specific types of bacteria present, the temperature of the flesh of the fish and of the surrounding atmosphere, and the pH of the fish. It is known that to extend the shelf life of fish, one may, for example, reduce the number of bacteria present using chemical means, freezing or other methods, create an acidic pH and/or maintain the product below 5° C. in its fresh state. The most common process employed to extend the shelf life of fish is freezing.

An inherent problem with freezing fish is its loss of the "fresh" attributes such as a "pink" or "red" meat colour to both the fish flesh and the "blood line" in the fish. The loss of these attributes causes the value of the frozen fish to be much less than the value of fish that has not been previously frozen. This loss of value is an interpretation of the quality of the fish by the consumer. The colour of the flesh and blood line of the fish is a major factor in the selling of seafood at the consumer level. Most consumers purchase fish with their "eyes" rather than with any other factor, such as smell, taste or texture. Therefore, it is desirable to maintain the "fresh" pink/red colour of the seafood products as long as possible in order to sell the product at a premium to consumers.

Although many factors may effect changes to the colour of fish products, the main reduction of colour results from damage to the hemoglobin pigments in the fish. Several of the primary causes for the reduction of hemoglobin pigments, resulting in a corresponding reduction in the "fresh" colour of the fish, include oxidation of the "red" hemoglobin pigments in the flesh to a "brown" colour; bacterial decomposition of the cells containing the hemoglobin pigments; and destruction and oxidation of the hemoglobin pigment during freezing.

Most unfrozen fish is considered "fresh" for as many as 30 days from catching. However, unfrozen fish this old usually contains high levels of dangerous bacterial decomposition. Bacterial decomposition of fish is the cellular breakdown of the flesh of the fish due to the digestive enzymes of bacteria present on or within the flesh of the fish. Conversely, frozen fish is usually frozen upon catching which reduces the likelihood that the fish will contain significant or harmful levels of bacterial decomposition.

In order to preserve the freshness of the fish and maintain the colour of the flesh and blood line to a satisfactory consumer level, processes using smoking and freezing techniques have been applied.

Smoking of fish has been one of the major forms of fish preservation for centuries. Smoking involves the burning of organic substances, such as wood, to produce a complex mix of over 400 separate chemical compounds. These compounds, when continually exposed to fish flesh, are absorbed into the meat over time and impart a smoke flavour to the flesh. The smoke compounds act as a natural "bacteriostat" and greatly increase the refrigerated shelf life of the flesh (up to three times the un-smoked shelf life). Smoking of fish increases the shelf life by killing a majority of the bacteria initially present, and then creating an acidic environment that slows the growth of bacteria over time in refrigerated conditions. The compounds in the smoke that are primarily responsible for the extension of the shelf life of fish are the aldehydes and phenols, as well as CO, CO_2, NO, NO_2, which are the main gaseous components of smoke. These compounds maintain the "fresh" colour of the fish, as well as prevent the growth of bacteria both on the surface of the fish and within the flesh.

However, one of the problems inherent in smoking fish products to impart preservation properties is that the smoke odour and/or smoke taste remains present in the fish flesh. Additionally, smoke that is produced from organic fuel materials typically contains particulates, such as creosote, tar, soot, etc., which are undesirable elements to have in contact with the fish product. Thus, it is beneficial to provide a smoke that has had some of the particulate removed and further remove the smoke odour/taste while still maintaining the extended shelf life.

A process for manufacturing a tasteless, super-purified smoke for the treatment of seafood and meat. The super-purified smoke is then applied to seafood or meat to preserve the freshness, colour, texture, and natural flavour, particularly after the seafood or meat is frozen and thawed. Kowalski teaches that the smoke must be super-purified by filtering out a substantial amount of odour and taste imparting particulate matter and gaseous vapors, thereby recovering the smoke in a tasteless form. Thus, Kowalski is limited in that it requires that the smoke be super-purified into a tasteless form in order to prevent the impartation of the smoke odour or taste to the seafood or meat products.

A process for smoking fish and meat at low temperatures, thereby conferring a smoked flavour and taste, and further preventing decomposition and discolouration of the fish or meat. As in Kowalski, the smoke is filtered to remove the larger particulates and provide a smoke that will preserve, sterilise and aid in maintaining the colour of the flesh of the fish or meat. However, Yamaoka teaches that the smoke odour or taste will remain in the fish or meat and that the temperature of application of the smoke is important. Specifically, the Yamaoka smoke preservation process must be carried out at extremely low temperatures (between 0 and 5° C.) in order to maintain the freshness and quality of the fish or meat products Therefore, Yamaoka is limited to a smoke process for preserving fish or meat products wherein the product will retain a smoke odour or taste, and the process is further limited to a narrow range of temperature conditions.

A method for partially drying and then smoking fish fillets to preserve them. The fish fillets were first dried to remove a substantial portion of the moisture present and then treated within a smoke atmosphere. This method imparted a smoke flavour to the dried fillets and aided in the prevention of the fish deterioration.

It is also known to preserve the freshness or colour of fish or other meat products by several other methods of treatment. U.S., Pat. No. 3,859,450 to Alsina teaches that melanosis (blackening) in shellfish is prevented by application of an innocuous acid solution followed by carbon dioxide gas. The resultant chemical reaction between the acid solution and the carbon dioxide produces carbonic anhydride that penetrates the shellfish and prevents melanosis during preservation by freezing. The process also discloses that the use of a food preservative, such as metabisulphite, will prolong the preservation of the original taste and texture of the shellfish after thawing.

A process for maintaining good colour in meat, poultry and fish products. Specifically, Woodruff teaches that subjecting the product to an atmosphere containing a low oxygen concentration and followed by an atmosphere containing a small amount of carbon monoxide will convert oxymyoglobin to carboxymyoglobin. The process produces a red colour in the product and permits lengthy refrigeration of the product (two to three weeks). Further preservation is accomplished by Woodruff by maintaining the product in a modified carbon dioxide atmosphere or by freezing.

The freshness of meat or fish may be improved by treatment with ubidecarenone to prevent discolouration of the product. The ubidecarenone additive prevents the oxidation of the haem pigments, thereby maintaining the red colour of "fresh" product by preventing discolouration to a brown or gray appearance.

Ozone, a GRAS (generally regarded as safe) substance, has been used for more than ten years to sanitise, deodorise and prevent bacterial growth in food items. Its main strength is in the killing of surface and subsurface bacteria that lead to decomposition of fish flesh during refrigerated storage. Ozone may be applied using a gaseous or liquid medium or a combination thereof.

A process of treating poultry with ozone and ozone dissolved in water to reduce the population of contaminating organisms. The product is first subjected to a solution containing ozone and then exposed to a gaseous atmosphere containing ozone. The product is also subjected intermittently to UV exposure which further acts as a bactericide and decomposes any ozone remaining on the product into oxygen.

Although, it is known that the foregoing techniques may be used to preserve the fish flesh itself, these techniques often result in an appearance of fish that has lost its "fresh" attributes. Accordingly, without the 'pink' or 'red' colour of the fish flesh, consumers often consider such preserved fish as "not fresh," resulting in a lower sales price for the fish. The foregoing techniques claim to maintain the colour of the fish do so with the addition of chemical additives and preservatives which can alter the taste and texture of the fish or be toxic in certain dosages to humans. Additionally, maintaining the "fresh" attributes of the fish is not taught when the fish is preserved or further preserved by freezing.

Therefore, a heretofore unaddressed need exists in the industry to satisfy the aforementioned deficiencies and inadequacies and provide a preserved fish that retains all of the qualities and characteristics of a "day caught" fish.

SEAFOOD PRESERVATION BY HURDLE TECHNOLOGY

Hurdle technology has been defined by Leistner (2000) as an intelligent combination of hurdles which secures the microbial safety and stability as well as the organoleptic and nutritional quality and the economic viability of food products. Hurdle technology that pathogens in food products can be eliminated or controlled. Each hurdle implies putting microorganisms in a hostile environment, which inhibits their growth or causes their death. Some of those hurdles have been empirically used for years to stabilise meat, fish, milk and vegetables.

Examples of hurdles in a food system are high temperature during processing, low temperature during storage, increasing the acidity, lowering the water activity or redox potential, or the presence of preservatives. According to the type of pathogens and how risky they are, the intensity of the hurdles can be adjusted individually to meet consumer preferences in an economical way, without compromising the safety of the product

HURDLES IN FOODS

The most important hurdles used in food preservation are temperature (high or low), water activity(aw), acidity (pH), redox potential (Eh), preservatives (*e.g.*, nitrite, sorbate, sulfite), and competitive microorganisms (*e.g.*, lactic acid bacteria). However, more than 60 potential hurdles for foods, which improve the stability and/or quality of the products have already been described, and the list of possible hurdles for food preservation is by no means complete. Some hurdles (*e.g.*, Maillard reaction products) will influence the safety and the quality of foods, because they have antimicrobial properties and at the same time improve the flavour of the products. The same hurdle could have a positive or a negative effect on foods, depending on its intensity. For instance, chilling to an unsuitable low temperature is detrimental to some foods of plant origin ('chilling injury'), whereas moderate chilling will be beneficial for their shelf life. Another example is the pH of fermented sausage which should be low enough to inhibit pathogenic bacteria, but not so low as to impair taste. If the intensity of a particular hurdle in a food is too small it should be strengthened, if it is detrimental to the food quality it should be lowered. By this adjustment, hurdles in foods can be kept in the optimal range, considering safety as well as quality, and thus the total quality of a food.

From an understanding of the hurdle effect, the hurdle technology has been derived, which means that hurdles are deliberately combined to improve the microbial stability and the sensory quality of foods as well as their nutritional and economic properties. Thus, hurdle technology aims to improve the total quality of foods by application of an intelligent mix of hurdles. Over the years the insight into the hurdle effect has been broadened and the application of hurdle technology was extended. Inindustrialised countries the hurdle technology approach is currently of most interest for minimally processed foods which are mildly heated or fermented, and for underpinning the microbial stability and safety of foods coming from future lines, *e.g.*, healthful foods with less fat and/or salt or advanced hurdle-technology foods requiring only minimal packaging. For refrigerated foods chill temperatures are the major and sometimes the only hurdle. However, if exposed to temperature abuse during distribution of the foods, this hurdle breaks down, and spoilage or food poisoning could happen. Therefore, additional hurdles should be incorporated as safeguards into chilled foods, using an approach called 'invisible technology'.

Basic Aspects

Food preservation implies putting microorganisms in a hostile environment, in order to inhibit their growth or shorten their survival or cause their death. The feasible responses of microorganisms to this hostile environment determine whether they may grow or die. More research is needed in view of these responses; however, recent advances have been made by considering the homeostasis, metabolic exhaustion, and stress reactions of microorganisms in exhaustion, and stress reactions of microorganisms inducing the novel concept of multitarget preservation for a gentle but most effective preservation of hurdle-technology foods.

PRESERVING FISH SAFELY: CANNING, FREEZING, PICKLING AND SMOKING

Freshly caught fish spoil easily and need to be properly preserved. The four most popular methods of fish preservation are freezing, canning, smoking, and pickling. Top quality fresh fish are essential for fish preservation. Of all flesh foods, fish is the most susceptible to tissue decomposition, development of rancidity, and microbial spoilage. Keep freshly caught fish alive as long as possible. A metal link bag will permit fish to remain alive longer in the water than a stringer. Spoilage and slime-producing bacteria are present on every fish and multiply rapidly on a dead fish held in warm surface water.

Fish begin to deteriorate as soon as they leave the water. To delay spoilage, clean the fish as soon as possible. Thorough cleaning of the body cavity and chilling of the fish will prevent spoilage. Fish spoilage occurs rapidly at summer temperatures; spoilage is slowed down as freezing temperatures are approached.

FREEZING FISH

This is the simplest, most convenient, and most highly recommended method of fish preservation.

A good quality frozen product requires the following:

- Careful handling of the fish after catching.
- Wrapping material or method that is airtight and prevents freezer burn and the development of undesirable flavours.
- A freezer storage temperature of 0° F or lower.

To Freeze Fish

- *Option 1:* Remove the guts and thoroughly clean the fish soon after catching. Prepare the fish as you would for table use. Cut large fish into steaks or fillets. Freeze small fish whole. Wrap the fish in heavy-duty freezer bags. Separate layers of fish with two thicknesses of packaging material for easier thawing. Store at 0° F or lower. When ready to use, thaw in the refrigerator.
- *Option 2:* Cut large fish into steaks or fillets. Small fish, such as sunfish and panfish, or small servings of fish can be frozen in ice. Place the fish in a shallow pan or water-tight container. Cover with ice water and place in the freezer until frozen (8-12 hours). Remove block from container, wrap, and store in freezer.

The storage life of good quality frozen fish held at 0° f or lower follows:

- Northern pike, trout, whitefish, smelt, lake herring, carp - four to six months
- Chinook salmon, coho salmon, white bass - five to eight months
- Walleye, bass, crappie, sunfish, yellow perch, blue gill - eight to twelve months

Canning Fish

Fish is a low acid food and can be processed safely only at temperatures reached in a pressure canner. Failure to heat process fish at 240° F or higher may allow spores of the dangerous heat-resistant bacteria, Clostridium botulinum, to survive, germinate, and grow. The poison produced by botulinum bacteria causes botulism, a deadly food poisoning. The addition of small amounts of vinegar, or packing fish in tomato juice or tomato paste, does not remove the requirement for heat processing fish in a pressure canner.

Use standard heat-tempered canning jars. All processing times in this publication are for 1-pint jars. Wide-mouth pint jars will be easier to fill than narrower ones. General USDA method for canning fish without sauce (including blue, mackerel, salmon, steelhead, trout, and other fatty fish except tuna) Clean and gut fish within 2 hours after catching. Keep cleaned fish on ice until ready to can. Note: Glass-like crystals of magnesium ammonium phosphate sometimes form in canned salmon. There is no way for the home canner to prevent these crystals from forming, but they usually dissolve when heated and are safe to eat. Procedure: Remove head, tail, fins and scales. Wash and remove all blood. Split fish lengthwise, if desired. Cut cleaned fish into 3½ inch lengths. Fill pint jars, skin side next to glass, leaving 1 inch headspace. Do not add liquids. Adjust lids and process.

Processing procedures for minnesota altitudes:

1. Dial-gauge Pressure Canner
   Pints-100 minutes 11 PSI
2. Weighted-gauge Pressure Canner
   Pints-100 minutes 15 PSI

Pickling

Pickling is an easy method of preserving fish. Pickled fish must be stored in the refrigerator at no higher than 40° F (refrigerator temperature), and for best flavour must be used within four to six weeks. Only a few species of fish are preserved commercially by pickling, but almost any type of fish may be pickled at home.Refrigerate the fish during all stages of the pickling process.

Ingredients for Pickled Fish

- Fish - Use only fresh, high quality fish.
- Water - Avoid hard water, as it causes off colour and flavours.
- Vinegar - Use distilled, white vinegar with an acetic acid content of at least 5 percent (50 grains means the same thing). This percentage of acetic acid is needed to stop bacterial growth.
- Salt - Use high grade, pure canning or pickling salt. It does not contain calcium or magnesium compounds which may cause off colour and flavours in pickled fish.
- Spices

General Method for Precooked Pickled Fish

Soak fish in a weak brine (1 cup salt to 1 gallon of water) for one hour.

Drain the fish; pack in heavy glass, crock, enamel, or plastic container in strong brine (2-½ cups salt to 1 gallon of water) for 12 hours in refrigerator.

Rinse the fish in cold water. Combine the following ingredients in a large pan or kettle.

This makes enough for 10 pounds of fish:

- ¼ oz bay leaves
- 2 T allspice
- 2 T mustard seed
- 1 T whole cloves
- 1 T pepper, ground
- 1-2 T hot, ground dried pepper
- ½ lb onions, sliced
- 2 qt distilled vinegar
- 5 c water (avoid hard water of high mineral content)

Bring to a boil, add fish, and simmer for 10 minutes until fish is easily pierced with a fork. Remove fish from liquid, place on a single layer on a flat pan. Refrigerate and cool quickly to prevent spoilage. Pack cold fish in clean glass jars, adding a few whole spices, a bay leaf, freshly sliced onions, and a slice of lemon. Strain the vinegar solution, bring to a boil, and pour into jars until fish is covered. Seal the jar immediately with two-part sealing lid, following the manufacturer's instructions. Pickled fish must be stored in the refrigerator as stated in general directions.

Caution: The Broad Fish Tapeworm

The broad fish tapeworm infection can be contracted by humans from eating raw or undercooked species of fish found in the Great Lakes area. Those who wish to prepare raw pickled fish should first freeze the fish at 0° F for 48 hours. The larvae of the broad fish tapeworm pass through smaller fish until they lodge as hatched small worms in the flesh of large carnivorous species of fish, like northern pike, walleye pike, sand pike, burbot, and yellow perch. This worm, if eaten by humans in its infective stage, can attach to the small intestine and grow to lengths of 10 to 30 feet.

The infective worms are destroyed readily either by cooking or freezing. Two recent outbreaks of this tapeworm in Minnesota were related to eating uncooked pickled pike. Those who wish to prepare raw pickled fish should first freeze the fish at 0° F for 48 hours.

Smoking Fish

Smoking has long been used as a means of temporarily preserving fish. The steps in the smoking process are necessary not only for safe preservation, but also to produce good flavour and aroma. Carp, suckers, buffalo catfish, salmon, trout, and chubs may be successfully smoked. A safe, high quality product can be produced using the following brining and smoking procedures. Certain steps in the brining and smoking process require careful attention.

Brining:

- Use correct amount of salt in the brine.
- Use enough brine for a given amount of fish.
- The temperature during brining must be no higher than 40° F.
- Use similar size and kinds of fish in the brine.

Smoking:

- There should be uniform heat treatment of all fish in the smoking chamber.
- Use freshly caught, dressed fish, whole or filleted. Wash fish thoroughly.

PREPARATION AND PRESERVATION OF ALABAMA SEAFOOD

Many species of seafood are harvested, both recreationally and commercially, from the waters of the Gulf of Mexico. The four most economically valuable forms landed in Alabama are shrimp, blue crabs, oysters, and fish. Whether home consumers catch their own seafood or purchase it at the seafood market, they often want to do much of the preparation themselves. They may also wish to store some or all of their seafood for extended periods. This brochure provides recommendations for proper home handling, preparation, and preservation of Alabama seafood.

SHRIMP

Preparation

Head the shrimp promptly; simply pinch the heads off and discard them. Removing the heads reduces the amount of storage space you need because the head accounts for 35 to 40 percent of a shrimp's body weight. Shrimp heads also contain more than 80 percent of the spoilage bacteria found in shrimp. So, headed shrimp are less likely to spoil than those with heads left on. After the head is removed, the final cleaning involves removing the shell, tail, and "sand vein" (if desired).

- To peel the shrimp, hold the tail in one hand and slip the thumb of the other hand under the shell between the swimmerettes. Lift off several segments of shell. Repeat, if necessary, removing all but the tail section.
- If the tail section is to be removed, squeeze the tail with the thumb and forefinger. Pull the shrimp meat with the other hand until it is released from the shell.
- The sand vein (usually black) runs along the upper curve of the shrimp's body. To remove it, make a 1/8 inch deep cut with a sharp knife along the length of the upper curve. Then rinse the vein away under cold running water.

Preservation

Wash and sanitise your hands, the kitchen sink, counter top, and any other surfaces that will come in contact with the shrimp. Dissolve 1 tablespoonful of liquid laundry bleach in 1 gallon of tap water for a simple, effective sanitising solution. Wash the shrimp thoroughly, using plenty of cool tap water. Leave the shells on the tail meat because the shells help reduce drying out (freezer burn) during frozen storage. If you want to eat the shrimp fresh, mix them with ice and store them in the refrigerator. Uncooked shrimp should not be kept in the refrigerator for more than 3 to 4 days.

To freeze shrimp in zip-top freezer bags:

- Place 1 pound of shrimp in a 1-quart zip-top freezer bag.
- Fill the bag with cool tap water. Lay the bag on its side and drain the water off until the bag is almost flattened against the shrimp.
- Quickly zip the bag shut and freeze.

To freeze shrimp in milk cartons:

- Thoroughly wash and sanitise milk cartons using the solution described above.
- Place 2 pounds of shrimp in a half-gallon waxed milk carton.
- Fill the carton with cool tap water to within 1 inch of the top.
- Fold the top over and freeze.
- After the contents have frozen, open the carton and add more water to cover any exposed shrimp. Then, fold the top over again, tape it closed, and freeze.

Shrimp frozen by these methods will keep for 4 to 6 months. Keep them solidly frozen. Do not thaw and refreeze them. Repeated freezing and thawing reduces the quality of the shrimp and provides a potential for spoilage. Thaw the shrimp carefully, either overnight in the refrigerator or under cold, running tap water, immediately before use.

BLUE CRABS

Preparation of Hard-Shell Crabs

Hard-shell blue crabs are sold either live or as cooked meat. When buying live crabs, make sure they show movement. Store the live crabs in a cool, moist area with plenty of air. Do not store them in airtight containers or containers filled with water because they will die. Crabs are normally very active, but if they are kept out of the water for long periods, they will slow their movements. Some will even die. To check crabs that are not moving, gently tap on the top shell with a stick or other utensil. Live crabs will respond by quickly raising their claws. Crabs that don't respond are dead and should be discarded immediately. Wash crabs thoroughly with plenty of cool tap water. This is best done with a kitchen sink sprayer or garden hose. Continue spraying the crabs until water that drains from the holding container is clear and trash free. To prepare whole live crabs, boil them in seasoned water. Bring the seasoned water to a boil in a large pot or kettle. Add washed live crabs and cover tightly. After the water resumes boiling, cook the crabs at a full, rolling boil for at least 15 minutes. Remove the crabs from the heat, drain, and allow them to cool. If you want to freeze the crabs, they will keep better if they are cooked first.

You can also clean the crabs before you cook them. This way, you can cook only the claws and the inner skeleton (core), which contains the white meat. In this case, the cooking time should be only 8 to 10 minutes.

To clean hard-shell crabs:

- With the crab upside down, firmly grasp the legs on one side with one hand and, with the other hand, lift the flap (apron) and pull back and down to remove the top shell.
- Turn the crab right side up and remove the gills. Wash out the intestines and spongy material.

To remove the meat:

- With a twisting motion, pull the legs loose from the body. Remove any meat that clings to the legs. Break off the claws.
- Slice off the top of the inner skeleton and remove all the exposed meat on this slice.
- At the back of the crab, on each side, lies a large lump of meat. With very careful U-shaped motions of the knife, remove these back fin lumps.
- Remove the white flake meat from the other pockets with the point of the knife.
- Crack the claw shell by striking the base of the first joint with a heavy, dull knife blade.

Remove the shell from around the movable pincer. This will expose the claw meat. If the meat is left attached to the movable pincer, it will make a delicious crab finger hors d'oeuvre. The dark meat can also be removed and used in soups, salads, or casseroles.

Preservation of Hard-Shell Crabs

Remove the spongy, yellowish-orange structures (digestive and reproductive organs) from the body cavity and discard. Also, remove and discard the greyish-white, feathery structures (gills) found on either side of the body cavity. What remains is the body pod (made of cartilage), which contains the edible white meat. If you plan to eat the crab meat fresh, pick it from the claws and body pod and place it in a cleaned and sanitised plastic storage container. Or, place the claws and body pods themselves in plastic storage bags. The packaged pods or picked meat may be stored on ice in the refrigerator for 2 to 3 days.

If the crab meat is to be frozen, it is better to leave it in the claws and body pods. Picked crab meat is more easily damaged by the formation of ice crystals and freezer burn. Place the claws and body pods in half-gallon waxed milk cartons that have been cleaned and sanitised. Add cool tap water to within 1 inch of the top. Fold the top over, tape it closed, and freeze. Cooked hard-shell blue crab meat does not hold up very well in the freezer. Crabs frozen by this method will keep for only about 1 month. Do not thaw and refreeze them. Thaw the crabs overnight in the refrigerator only. Thawing under running tap water washes away the flavour.

Preparation of Soft-Shell Crabs

Crabs periodically shed their shells so they can grow. A new, soft shell forms beneath the old, hard one. The crab backs out of the old shell as it loosens and splits. The crab then absorbs water to expand to its new, larger size. Soft-shell blue crabs are available live, fresh, or frozen and either whole or cleaned. When buying live soft-shell blue crabs, make sure they show movement. Store live crabs in a cool, moist area with plenty of air, not in airtight containers or small containers filled with water. For maximum quality, use them within 24 hours.

Clean live soft-shell blue crabs or whole (round) fresh or frozen crabs before cooking.

Follow these easy steps:

- Lift the large top shell spines (points) and scrape out the gills (grayish white feathery structures).
- Remove the eyes and mouth parts by cutting across the front of the crab just behind the eyes with a knife or a pair of scissors.
- Press the top shell over the legs to remove the bile sacs. Scrape out the internal organs through the cut made to remove the face. Rinse the body cavity with cold water. Or, the entire top shell can be lifted to expose the internal organs. Remove the internal organs and rinse the body cavity with cold water.
- Turn the crab over and lift and remove the apron at its base.

Preservation of Soft-Shell Crabs

Fresh soft-shell crabs will maintain their quality better when wrapped in plastic and packed in ice in the refrigerator. For maximum quality, use them within 48 hours.

To freeze soft-shell blue crabs, wrap them in several layers of moisture-proof plastic wrap and place them in the freezer. Properly wrapped and hard frozen soft-shell blue crabs will maintain good quality for up to 6 months. Do not thaw and refreeze them. Thaw the crabs overnight in the refrigerator only.Thawing under running tap water washes away the flavour.

OYSTERS

Preparation

Thoroughly wash and scrub all mud and debris from the shells. This is best done with a garden hose and scrub brush. Make sure that the oysters are alive. Live oysters hold their shells tightly closed. Tap any oysters with slightly opened or "gaped" shells. Live oysters will respond by closing their shells tightly. Unresponsive oysters are dead and should be discarded. To shuck whole oysters, use cotton gloves and an oyster knife, which has a heavy, wedge-shaped blade and handle, often made in one piece. It is designed to withstand the pressure required to open oysters. Never use a sharp knife.

The cleanest way to open oysters is to grasp the oyster securely by the thin end or "bill," leaving the hinge (thicker portion) exposed towards the other hand.

Then:

- Insert the oyster knife in the crevice between the shells at the hinge; twist the knife while pushing it firmly into the opening to sever the hinge.
- Once the hinge is broken, before pulling the shell apart, slide the knife along the inside of the top shell to cut the adductor muscle loose from the shell.
- Remove the top shell and slip the knife under the body of the oyster, being careful not to mutilate it, and cut the muscle away from the bottom shell. Remove any remaining shell particles that may be attached to the oyster. Most oysters, except the very largest, can be opened by this method.

Preservation

Store oysters to be eaten fresh in the refrigerator. Unshucked oysters can be stored without ice in the refrigerator and should remain alive for 7 to 10 days. Shucked oyster meats may be placed in sanitised 1-cup or 1-pint containers or plastic bags, packed in ice, and placed in the refrigerator. Oysters stored by this method will keep for 7 to 10 days.

Frozen oysters will be of lesser quality than fresh oysters. The simplest method is to freeze oysters in the shell. Place the oysters in a plastic bag and press out excess air. Seal the bag and freeze. The shell and juices provide an excellent, natural container for the oyster meat. Shucked oyster meats can be frozen also. They will maintain a better flavour if frozen in their own natural juices (liquor). Place shucked oyster meats insanitised plastic containers or bags. Press the bags flat. Leave 1/2 inch of headspace in the containers. Freeze as quickly as possible. Oysters frozen by either of these methods will keep for 2 to 3 months. Do not thaw and refreeze the oysters. Thaw overnight in refrigerator only. Thawing under running water washes away the flavour.

FISH

Preparation

Wash and sanitise your hands, all work areas, and storage containers. Before cooking or storing your fish, dress it to the form that best suits your needs.

To prepare a whole (round) fish for cooking or storage, follow these simple steps:

- *Scale:* Lay the fish on a board and hold the tail. Working from the tail to the head, scrape the scales off using a scaler, stiff knife, or large spoon. Rinse well.
- *Eviscerate:* Make a cut along the entire length of the belly of the fish. Remove the entrails and pelvic fins. Clean the belly cavity thoroughly under running water.
- *Dress:* Remove the head and pectoral fins by making a cut just behind them with a sharp knife. If the backbone is large, cut down to it on both sides and snap the head off. Remove the dorsal fin by cutting along each side with a sharp knife. Grasp the end near the tail and give a quick pull towards the head.
- *Steak:* A large dressed fish can be cut into steaks by cutting across the backbone at 1- to 1-1/4-inch intervals.
- *Fillet:* For filleting, it is not necessary to completely dress the fish. Scale the fish, unless the fillet is to be skinned. Use a sharp, flexible, thin-blade knife. Cut through the flesh along the backbone from the tail to just behind the head. Cut down to the collar bone. Turn the knife flat and slide it along the ribs to the tail, cutting the flesh away from the backbone. Repeat on the other side.
- *Rinse:* Thoroughly wash the dressed fish, steaks, or fillets under plenty of cool, running tap water.

Many large fish can have tapeworms, which are usually found in the tail quarter of the fish. To remove them, fillet the fish in the usual fashion. The affected area will have a reddish tinge, unlike the whitish, unaffected areas. The tapeworms themselves are white. Although there should be no danger from eating this portion of the fish, most people will probably want to remove and discard the affected section.

Preservation

If you plan to eat your fish fresh, wrap it in clear plastic or place it in zip-top storage bags. Pack the fish in ice and place it in the refrigerator.

Fresh fish stored in this manner will keep for 5 to 7 days. To freeze fish follow the directions for freezing shrimp.

To glaze fish:

- Dip each fish portion or fillet in ice water. Lay them on a cookie sheet (not touching) and place in freezer.
- After the fish are solidly frozen, dip them in ice water again, place them back on the cookie sheet, and return to freezer.
- Repeat Step 2 several times until an ice glaze completely covers the fish portions.
- Wrap the glazed fish in two layers of plastic wrap or seal them in plastic bags.
- Place fish in freezer.

Most fish frozen by these methods will keep for 4 to 6 months. However, fish with high fat content, such as mackerel, mullet, or bluefish, maintain their quality while frozen for only about 3 months. Keep the fish solidly frozen. Do not thaw and refreeze them. Repeated thawing and refreezing reduces the quality of the fish and provides a potential for spoilage. Thaw fish carefully, either overnight in the refrigerator or under cold, running tap water immediately before use.

NUTRITION

Nutrition (also called nourishment or aliment) is the provision, to cells and organisms, of the materials necessary (in the form of food) to support life. Many common health problems can be prevented or alleviated with a healthy diet. The diet of an organism is what it eats, which is largely determined by the perceived palatability of foods. Dietitians are health professionals who specialize in human nutrition, meal planning, economics, and preparation.

They are trained to provide safe, evidence-based dietary advice and management to individuals (in health and disease), as well as to institutions. A poor diet can have an injurious impact on health, causing deficiency diseases such as scurvy, beriberi, and kwashiorkor; health-threatening conditions like obesity and metabolic syndrome; and such common chronic systemic diseases as cardiovascular disease, diabetes, and osteoporosis.

Nutrition science investigates the metabolic and physiological responses of the body to diet. With advances in the fields of molecular biology, biochemistry, and genetics, the study of nutrition is increasingly concerned with metabolism and metabolic pathways: the sequences of biochemical steps through which substances in living things change from one form to another. Nitrogen is needed by animals to build proteins. Carnivore and herbivore diets vary in their source of nitrogen, which is a limiting nutrient for both. Herbivores consume plants to get nitrogen and carnivores consume other animals to obtain nitrogen. Nitrogen is a common element in the atmosphere but exists in a state that is not usable by most living organisms, certain fungi and bacteria are able to convert atmospheric nitrogen into a form plants can adsorb and utilize.

The human body contains chemical compounds, such as water, carbohydrates (sugar, starch, and fibre), amino acids (in proteins), fatty acids (in lipids), and nucleic acids (DNA and RNA). These compounds in turn consist of elements such as carbon, hydrogen, oxygen, nitrogen, phosphorus, calcium, iron, zinc, magnesium, manganese, and so on. All of these chemical compounds and elements occur in various forms and combinations (e.g., hormones, vitamins, phospholipids, hydroxyapatite), both in the human body and in the plant and animal organisms that humans eat.

The human body consists of elements and compounds ingested, digested, absorbed, and circulated through the bloodstream to feed the cells of the body. Except in the unborn foetus, which receive processed nutrients from the mother, the digestive system is the first system involved in breaking down food prior to further digestion.

Digestive juices, excreted into the lumen of the gastrointestinal tract, break chemical bonds in ingested molecules, and modulate their conformations and energy states. Though some molecules are absorbed into the bloodstream unchanged, digestive processes release them from the matrix of foods. Unabsorbed matter, along with some waste products of metabolism, is eliminated from the body in the feces. Studies of nutritional status must take into account the state of the body before and after experiments, as well as the chemical composition of the whole diet and of all material excreted and eliminated from the body (in urine and foeces).

Comparing the food to the waste can help determine the specific compounds and elements absorbed and metabolized in the body. The effects of nutrients may only be discernible over an extended period, during which all food and waste must be analysed. The number of variables involved in such experiments is high, making nutritional studies time-consuming and expensive, which explains why the science of human nutrition is still slowly evolving. In general, eating a wide variety of fresh, whole (unprocessed), foods has proven favourable for one's health compared to monotonous diets based on processed foods. In particular, the consumption of whole-plant foods slows digestion and allows better absorption, and a more favourable balance of essential nutrients per Calorie, resulting in better management of cell growth, maintenance, and mitosis (cell division), as well as better regulation of appetite and blood sugar. Regularly scheduled meals (every few hours) have also proven more wholesome than infrequent or haphazard ones, although a recent study has also linked more frequent meals with a higher risk of colon cancer in men.

There are six major classes of nutrients: carbohydrates, fats, minerals, protein, vitamins, and water. These nutrient classes can be categorized as either macronutrients (needed in relatively large amounts) or micronutrients (needed in smaller quantities). The macronutrients include carbohydrates, fats, protein, and water. The micronutrients are minerals and vitamins. The macronutrients (excluding water) provide structural material (amino acids from which proteins are built, and lipids from which cell membranes and some signaling molecules are built), energy.

Some of the structural material can be used to generate energy internally, and in either case it is measured in Joules or kilocalories (often called "Calories" and written with a capital C to distinguish them from little 'c' calories). Carbohydrates and proteins provide 17 kJ approximately (4 kcal) of energy per gram, while fats provide 37 kJ (9 kcal) per gm., though the net energy from either depends on such factors as absorption and digestive effort, which vary substantially from instance to instance.

Vitamins, minerals, fibre, and water do not provide energy, but are required for other reasons. A third class of dietary material, fibre (*i.e.*, non-digestible material such as cellulose), is also required, for both mechanical and biochemical reasons, although the exact reasons remain unclear.

Molecules of carbohydrates and fats consist of carbon, hydrogen, and oxygen atoms. Carbohydrates range from simple monosaccharides (glucose, fructose, galactose) to complex polysaccharides (starch). Fats are triglycerides, made of assorted fatty acid monomers bound to glycerol backbone. Some fatty acids, but not all, are essential in the diet: they cannot be synthesized in the body. Protein molecules contain nitrogen atoms in addition to carbon, oxygen, and hydrogen. The fundamental components of protein are nitrogen-containing amino acids, some of which are essential in the sense that humans cannot make them internally. Some of the amino acids are convertible (with the expenditure of energy) to glucose and can be used for energy production just as ordinary glucose in a process known as gluconeogenesis.

By breaking down existing protein, some glucose can be produced internally; the remaining amino acids are discarded, primarily as urea in urine. This occurs normally only during prolonged starvation. Other micronutrients include antioxidants and phytochemicals, which are said to influence (or protect) some body systems. Their necessity is not as well established as in the case of, for instance, vitamins.

Most foods contain a mix of some or all of the nutrient classes, together with other substances, such as toxins of various sorts. Some nutrients can be stored internally (e.g., the fat soluble vitamins), while others are required more or less continuously. Poor health can be caused by a lack of required nutrients or, in extreme cases, too much of a required nutrient. For example, both salt and water (both absolutely required) will cause illness or even death in excessive amounts.

CARBOHYDRATES

Carbohydrates include sugars, starches and fibre. They constitute a large part of foods such as rice, noodles, bread, and other grain-based products.

Carbohydrates may be classified chemically as monosaccharides, disaccharides, or polysaccharides depending on the number of monomer (saccharide or sugar) units they contain.

Monosaccharides, disaccharides, and polysaccharides contain one, two, and three or more sugar units, respectively. Polysaccharides are often referred to as complex carbohydrates because they consist of long, sometimes branched chains of single sugar units. Mono- and disaccharides are called simple carbohydrates. Dietary advice frequently but erroneously suggests that complex carbohydrates are superior to simple because they take longer to digest and absorb. Simple carbohydrates, on the other hand, are said to cause a spike in blood glucose levels rapidly after ingestion. These traditional claims are false. In fact, many digestible polysaccharides are processed as rapidly and simple sugars in the human body.

On the other hand, some simple carbohydrates (fructose, for example) are processed in a different way and do not spike blood sugar. Thus the distinction between "complex" and "simple" does not predict the nutritional value or impact of carbohydrates. A better way of determining what effect particular foods may have on blood sugar and ultimately on health in general is the glycemic index.

Carbohydrates are not essential nutrients (with the likely exception of fibre), but are typically an important part of the human diet. While it would not be accurate to categorize all carbohydrates as "bad" nutritionally, some carbohydrate sources may well have deleterious effects on health, especially when consumed in large quantities. Highly processed carbohydrates (sugars and starches) as well as fructose consumed in large quantities have been implicated in negative.

FIBRE

Dietary fibre is a carbohydrate (or a polysaccharide) that is incompletely absorbed in humans and in some animals. Like all carbohydrates, when it is metabolized it can produce four Calories (kilocalories) of energy per gm. However, in most circumstances it accounts for less than that because of its limited absorption and digestibility. Dietary fibre consists mainly of cellulose, a large carbohydrate polymer that is indigestible because humans do not have the required enzymes to disassemble it.

There are two subcategories: soluble and insoluble fibre. Whole grains, fruits (especially plums, prunes, and figs), and vegetables are good sources of dietary fibre. There are many health benefits of a high-fibre diet. Dietary fibre helps reduce the chance of gastrointestinal problems such as constipation and diarrhoea by increasing the weight and size of stool and softening it. Insoluble fibre, found in whole-wheat flour, nuts and vegetables, especially stimulates peristalsis—the rhythmic muscular contractions of the intestines which move digesta along the digestive tract. Soluble fibre, found in oats, peas, beans, and many fruits, dissolves in water in the intestinal tract to produce a gel which slows the movement of food through the intestines. This may help lower blood glucose levels because it can slow the absorption of sugar. Additionally, fibre, perhaps especially that from whole grains, is thought to possibly help lessen insulin spikes, and therefore reduce the risk of type 2 diabetes. The link between increased fibre consumption and a decreased risk of colorectal cancer is still uncertain.

FAT

A molecule of dietary fat typically consists of several fatty acids (containing long chains of carbon and hydrogen atoms), bonded to a glycerol. They are typically found as triglycerides (three fatty acids attached to one glycerol backbone). Fats may be classified as saturated or unsaturated depending on the detailed structure of the fatty acids involved.

Saturated fats have all of the carbon atoms in their fatty acid chains bonded to hydrogen atoms, whereas unsaturated fats have some of these carbon atoms double-bonded, so their molecules have relatively fewer hydrogen atoms than a saturated fatty acid of the same length. Unsaturated fats may be further classified as monounsaturated (one double-bond) or polyunsaturated (many double-bonds).

Furthermore, depending on the location of the double-bond in the fatty acid chain, unsaturated fatty acids are classified as omega-3 or omega-6 fatty acids. Trans fats are a type of unsaturated fat with trans-isomer bonds; these are rare in nature and in foods from natural sources; they are typically created in an industrial process called (partial) hydrogenation. There are nine kilocalories in each gm. of fat. Saturated fats (typically from animal sources) have been a staple in many world cultures for millennia. Unsaturated fats (e. g., vegetable oil) are considered healthier while trans fats are to be avoided.

Saturated and some trans fats are typically solid at room temperature (such as butter or lard), while unsaturated fats are typically liquids (such as olive oil or flaxseed oil). Trans fats are very rare in nature, and have been shown to be highly detrimental to human health, but have properties useful in the food processing industry, such as rancidity resistance.

ESSENTIAL FATTY ACIDS

Most fatty acids are non-essential, meaning the body can produce them as needed, generally from other fatty acids and always by expending energy to do so. However, in humans, at least two fatty acids are essential and must be included in the diet. An appropriate balance of essential fatty acids—omega-3 and omega-6 fatty acids—seems also important for health, although definitive experimental demonstration has been elusive. Both of these "omega" long-chain polyunsaturated fatty acids are substrates for a class of eicosanoids known as prostaglandins, which have roles throughout the human body. They are hormones, in some respects.

The omega-3 eicosapentaenoic acid (EPA), which can be made in the human body from the omega-3 essential fatty acid alpha-linolenic acid (LNA), or taken in through marine food sources, serves as a building block for series 3 prostaglandins (e.g., weakly inflammatory PGE3). The omega-6 dihomo-gamma-linolenic acid (DGLA) serves as a building block for series 1 prostaglandins (*e.g.*, anti-inflammatory PGE1), whereas arachidonic acid (AA) serves as a building block for series 2 prostaglandins (e.g., pro-inflammatory PGE 2). Both DGLA and AA can be made from the omega-6 linoleic acid (LA) in the human body, or can be taken in directly through food.

An appropriately balanced intake of omega-3 and omega-6 partly determines the relative production of different prostaglandins, which is one reason why a balance between omega-3 and omega-6 is believed important for cardiovascular health. In industrialized societies, people typically consume large amounts of processed vegetable oils, which have reduced amounts of the essential fatty acids along with too much of omega-6 fatty acids relative to omega-3 fatty acids. The conversion rate of omega-6 DGLA to AA largely determines the production of the prostaglandins PGE1 and PGE2.

Omega-3 EPA prevents AA from being released from membranes, thereby skewing prostaglandin balance away from pro-inflammatory PGE2 (made from AA) towards anti-inflammatory PGE1 (made from DGLA). Moreover, the conversion (desaturation) of DGLA to AA is controlled by the enzyme delta-5-desaturase, which in turn is controlled by hormones such as insulin (up-regulation) and glucagon (down-regulation).

The amount and type of carbohydrates consumed, along with some types of amino acid, can influence processes involving insulin, glucagon, and other hormones; therefore the ratio of omega-3 versus omega-6 has wide effects on general health, and specific effects on immune function and inflammation, and mitosis (*i.e.*, cell division).

PROTEIN

Proteins are the basis of many animal body structures (e.g., muscles, skin, and hair). They also form the enzymes that control chemical reactions throughout the body. Each molecule is composed of amino acids, which are characterized by inclusion of nitrogen and sometimes sulphur (these components are responsible for the distinctive smell of burning protein, such as the keratin in hair).

The body requires amino acids to produce new proteins (protein retention) and to replace damaged proteins (maintenance). As there is no protein or amino acid storage provision, amino acids must be present in the diet. Excess amino acids are discarded, typically in the urine. For all animals, some amino acids are essential (an animal cannot produce them internally) and some are non-essential (the animal can produce them from other nitrogen-containing compounds). Twenty-one proteinogenic amino acids are found in the human body, along with non-proteinogenic amino acids (e.g., gamma-aminobutyric acid).

Ten of the proteinogenic amino acids are essential and, therefore, must be included in the diet. A diet that contains adequate amounts of amino acids (especially those that are essential) is particularly important in some situations: during early development and maturation, pregnancy, lactation, or injury (a burn, for instance). A complete protein source contains all the essential amino acids; an incomplete protein source lacks one or more of the essential amino acids.

It is possible to combine two incomplete protein sources (e.g., rice and beans) to make a complete protein source, and characteristic combinations are the basis of distinct cultural cooking traditions. Sources of dietary protein include meats, tofu and other soy-products, eggs, legumes, and dairy products such as milk and cheese. Excess amino acids from protein can be converted into glucose and used for fuel through a process called gluconeogenesis. The amino acids remaining after such conversion are discarded.

MINERALS

Dietary minerals are the chemical elements required by living organisms, other than the four elements carbon, hydrogen, nitrogen, and oxygen that are present in nearly all organic molecules. The term "mineral" is archaic, since the intent is to describe simply the less common elements in the diet. Some are heavier than the four just mentioned, including several metals, which often occur as ions in the body. Some dietitians recommend that these be supplied from foods in which they occur naturally, or at least as complex compounds, or sometimes even from natural inorganic sources (such as calcium carbonate from ground oyster shells). Some minerals are absorbed much more readily in the ionic forms found in such sources. On the other hand, minerals are often artificially added to the diet as supplements; the most famous is likely iodine in iodized salt which prevents goiter.

MACROMINERALS

Many elements are essential in relative quantity; they are usually called "bulk minerals". Some are structural, but many play a role as electrolytes. Elements with recommended dietary allowance (RDA) greater than 200 mg/day are, in alphabetical order (with informal or folk-medicine perspectives in parentheses):

- Calcium, a common electrolyte, but also needed structurally (for muscle and digestive system health, bone strength, some forms neutralize acidity, may help clear toxins, provides signaling ions for nerve and membrane functions).

- Chlorine as chloride ions; very common electrolyte;
- Magnesium, required for processing ATP and related reactions (builds bone, causes strong peristalsis, increases flexibility, increases alkalinity).
- Phosphorus, required component of bones; essential for energy processing.
- Potassium, a very common electrolyte (heart and nerve health).
- Sodium, a very common electrolyte; not generally found in dietary supplements, despite being needed in large quantities, because the ion is very common in food: typically as sodium chloride, or common salt. Excessive sodium consumption can deplete calcium and magnesium, leading to high blood pressure and osteoporosis (Note: Some sources suggest high blood pressure is due to high water retention per osmosis).
- Sulfur, for three essential amino acids and therefore many proteins (skin, hair, nails, liver, and pancreas). Sulfur is not consumed alone, but in the form of sulfur-containing amino acids.

TRACE MINERALS

Many elements are required in trace amounts, usually because they play a catalytic role in enzymes.

Some trace mineral elements (RDA < 200 mg/day) are, in alphabetical order:

- Cobalt required for biosynthesis of vitamin B_{12} family of coenzymes. Animals cannot biosynthesize B_{12}, and must obtain this cobalt-containing vitamin in the diet.
- Copper required component of many redox enzymes, including cytochrome c oxidase.
- Chromium required for sugar metabolism.
- Iodine required not only for the biosynthesis of thyroxine, but probably, for other important organs as breast, stomach, salivary glands, thymus, etc.; for this reason iodine is needed in larger quantities than others in this list, and sometimes classified with the macrominerals.
- Iron required for many enzymes, and for haemoglobin and some other proteins.
- Manganese (processing of oxygen).
- Molybdenum required for xanthine oxidase and related oxidases.
- Nickel present in urease.
- Selenium required for peroxidase (antioxidant proteins).
- *Vanadium (Speculative*: there is no established RDA for vanadium. No specific biochemical function has been identified for it in humans, although vanadium is required for some lower organisms.)

- Zinc required for several enzymes such as carboxypeptidase, liver alcohol dehydrogenase, and carbonic anhydrase.

VITAMINS

Some vitamins are recognized as essential nutrients, necessary in the diet for good health. (Vitamin D is the exception: it can be synthesized in the skin, in the presence of UVB radiation.) Certain vitamin-like compounds that are recommended in the diet, such as carnitine, are thought useful for survival and health, but these are not "essential" dietary nutrients because the human body has some capacity to produce them from other compounds.

Moreover, thousands of different phytochemicals have recently been discovered in food (particularly in fresh vegetables), which may have desirable properties including antioxidant activity, however, experimental demonstration has been suggestive but inconclusive. Other essential nutrients that are not classified as vitamins include essential amino acids, choline, essential fatty acids, and the minerals discussed in the preceding part.

Vitamin deficiencies may result in disease conditions, including goitre, scurvy, osteoporosis, impaired immune system, disorders of cell metabolism, certain forms of cancer, symptoms of premature aging, and poor psychological health (including eating disorders), among many others. Excess levels of some vitamins are also dangerous to health (notably vitamin A), and for at least one vitamin, B6, toxicity begins at levels not far the required amount. Deficient or excess levels of minerals can also have serious health consequences.

WATER

It is not fully clear how much water intake is needed by healthy people, although some assert that 6–8 glasses of water daily is the minimum to maintain proper hydration. The notion that a person should consume eight glasses of water per day cannot be traced to a credible scientific source. The effect of, greater or lesser, water intake on weight loss and on constipation is also still unclear. The original water intake recommendation in 1945 by the Food and Nutrition Board of the National Research Council read: "An ordinary standard for diverse persons is 1 milliliter for each calorie of food.

Most of this quantity is contained in prepared foods." The latest dietary reference intake report by the United States National Research Council recommended, generally, (including food sources): 2.7 litres of water total for women and 3.7 litres for men. Specifically, pregnant and breastfeeding women need additional fluids to stay hydrated. According to the Institute of Medicine—who recommend that, on average, women consume 2.2 litres and men 3.0 litres—this is recommended to be 2.4 litres (approx. 9 cups) for pregnant women and 3 litres (approx. 12.5 cups) for breastfeeding women because an especially large amount of fluid is lost during nursing.

For those who have healthy kidneys, it is somewhat difficult to drink too much water, but (especially in warm humid weather and while exercising) it is dangerous to drink too little. People can drink far more water than necessary while exercising, however, putting them at risk of water intoxication, which can be fatal. In particular, large amounts of de-ionized water are dangerous. Normally, about 20 per cent of water intake comes in food, while the rest comes from drinking water and assorted beverages (caffeinated included). Water is excreted from the body in multiple forms; including urine and faeces, sweating, and by water vapour in the exhaled breath.

ANTIOXIDANTS

As cellular metabolism/energy production requires oxygen, potentially damaging (e.g., mutation causing) compounds known as free radicals can form. Most of these are oxidizers (*i.e.*, acceptors of electrons) and some react very strongly. For the continued normal cellular maintenance, growth, and division, these free radicals must be sufficiently neutralized by antioxidant compounds.

Some are produced by the human body with adequate precursors (glutathione, Vitamin C), and those the body cannot produce may only be obtained in the diet via direct sources (Vitamin C in humans, Vitamin A, Vitamin K) or produced by the body from other compounds (Beta-carotene converted to Vitamin A by the body, Vitamin D synthesized from cholesterol by sunlight). Phytochemicals and their subgroup, polyphenols, make up the majority of antioxidants; about 4,000 are known. Different antioxidants are now known to function in a cooperative network. For example, Vitamin C can reactivate free radical-containing glutathione or Vitamin E by accepting the free radical itself. Some antioxidants are more effective than others at neutralizing different free radicals.

Some cannot neutralize certain free radicals. Some cannot be present in certain areas of free radical development (Vitamin A is fat-soluble and protects fat areas, Vitamin C is water soluble and protects those areas).

When interacting with a free radical, some antioxidants produce a different free radical compound that is less dangerous or more dangerous than the previous compound.

Having a variety of antioxidants allows any byproducts to be safely dealt with by more efficient antioxidants in neutralizing a free radical's butterfly effect. Although initial studies suggested that antioxidant supplements might promote health, later large clinical trials did not detect any benefit and suggested instead that excess supplementation may be harmful.

PHYTOCHEMICALS

A growing area of interest is the effect upon human health of trace chemicals, collectively called phytochemicals. These nutrients are typically found in edible plants, especially colourful fruits and vegetables, but also other organisms including seafood, algae, and fungi. The effects of phytochemicals increasingly survive rigorous testing by prominent health organizations. One of the principal classes of phytochemicals are polyphenol antioxidants, chemicals that are known to provide certain health benefits to the cardiovascular system and immune system.

These chemicals are known to down-regulate the formation of reactive oxygen species, key chemicals in cardiovascular disease. Perhaps the most rigorously tested phytochemical is zeaxanthin, a yellow-pigmented carotenoid present in many yellow and orange fruits and vegetables. Repeated studies have shown a strong correlation between ingestion of zeaxanthin and the prevention and treatment of age-related macular degeneration (AMD).

Less rigorous studies have proposed a correlation between zeaxanthin intake and cataracts. A second carotenoid, lutein, has also been shown to lower the risk of contracting AMD. Both compounds have been observed to collect in the retina when ingested orally, and they serve to protect the rods and cones against the destructive effects of light.

Another carotenoid, beta-cryptoxanthin, appears to protect against chronic joint inflammatory diseases, such as arthritis.

While the association between serum blood levels of beta-cryptoxanthin and substantially decreased joint disease has been established, neither a convincing mechanism for such protection nor a cause-and-effect have been rigorously studied. Similarly, a red phytochemical, lycopene, has substantial credible evidence of negative association with development of prostate cancer.

Some of the correlations between the ingestion of certain phytochemicals and the prevention of disease are, in some cases, enormous in magnitude. Yet, even when the evidence is obtained, translating it to practical dietary advice can be difficult and counter-intuitive.

Lutein, for example, occurs in many yellow and orange fruits and vegetables and protects the eyes against various diseases. However, it does not protect the eye nearly as well as zeaxanthin, and the presence of lutein in the retina will prevent zeaxanthin uptake.

Additionally, evidence has shown that the lutein present in egg yolk is more readily absorbed than the lutein from vegetable sources, possibly because of fat solubility.

At the most basic level, the question "should you eat eggs?" is complex to the point of dismay, including misperceptions about the health effects of cholesterol in egg yolk, and its saturated fat content.

As another example, lycopene is prevalent in tomatoes (and actually is the chemical that gives tomatoes their red colour). It is more highly concentrated, however, in processed tomato products such as commercial pasta sauce, or tomato soup, than in fresh "healthy" tomatoes.

Yet, such sauces tend to have high amounts of salt, sugar, other substances a person may wish or even need to avoid.

The following table presents phytochemical groups and common sources, arranged by family:

Family	Sources	Possible Benefits
Flavonoids	Berries, herbs, vegetables, wine, grapes, tea	General antioxidant, oxidation of LDLs, prevention of arteriosclerosis and heart disease
Isoflavones (phytoestrogens)	Soy, red clover, kudzu root	General antioxidant, prevention of arteriosclerosis and heart disease, easing symptoms of menopause, cancer prevention
Isothiocyanates	Cruciferous vegetables	cancer prevention
monoterpenes	Citrus peels, essential oils, herbs, spices, green plants, atmosphere	Cancer prevention, treating gallstones
Organosulfur compounds	Chives, garlic, onions	Cancer prevention, lowered LDLs, assistance to the immune system
Saponins	Beans, cereals, herbs	Hypercholesterolemia, Hyperglycemia, Antioxidant, cancer prevention, Anti-inflammatory
Capsaicinoids	All capiscum (chile) peppers	Topical pain relief, cancer prevention, cancer cell apoptosis

INTESTINAL BACTERIAL FLORA

It is now also known that animal intestines contain a large population of gut flora. In humans, these include species such as Bacteroides, *L. acidophilus*, and *E. coli*, among many others. They are essential to digestion, and are also affected by the food we eat. Bacteria in the gut perform many important functions for humans, including breaking down and aiding in the absorption of otherwise indigestible food; stimulating cell growth; repressing the growth of harmful bacteria, training the immune system to respond only to pathogens; producing vitamin B_{12}, and defending against some infectious diseases.

5

Seafood and Peptidoglycan

Peptidoglycan, also known as murein, is a polymer consisting of sugars and amino acids that forms a mesh-like layer outside the plasma membrane of eubacteria. The sugar component consists of alternating residues of β-(1,4) linked N-acetylglucosamine and N-acetylmuramic acid residues. Attached to the N-acetylmuramic acid is a peptide chain of three to five amino acids.

The peptide chain can be cross-linked to the peptide chain of another strand forming the 3D mesh-like layer. Some Archaea have a similar layer of pseudopeptido-glycan. Peptidoglycan serves a structural role in the bacterial cell wall, giving structural strength, as well as counteracting the osmotic pressure of the cytoplasm. A common misconception is that peptidoglycan gives the cell its shape; however, whereas peptidoglycan helps maintain the structure of the cell, it is actually the MreB protein that facilitates cell shape. Peptidoglycan is also involved in binary fission during bacterial cell reproduction.

The peptidoglycan layer is substantially thicker in Gram-positive bacteria (20 to 80 nm) than in Gram-negative bacteria (7 to 8 nm), with the attachment of the S-layer. Peptidoglycan forms around 90% of the dry weight of Gram-positive bacteria but only 10% of Gram-negative strains. In Gram-positive strains, it is important in attachment roles and sterotyping purposes. For both Gram-positive and Gram-negative bacteria, particles of approximately 2 nm can pass through the peptidoglycan.

ANTIBIOTIC INHIBITION

Some antibacterial drugs such as penicillin interfere with the production of peptidoglycan by binding to bacterial enzymes known as penicillin-binding proteins or transpeptidases. Penicillin-binding proteins form the bonds between oligopeptide crosslinks in peptidoglycan.

For a bacterial cell to reproduce through binary fission, more than a million peptidoglycan subunits (NAM-NAG+oligopeptide) must be attached to existing subunits. Mutations in transpeptidases that lead to reduced interactions with an antibiotic are a significant source of emerging antibiotic resistance. Considered the human body's *own antibiotic*, lysozymes found in tears work by breaking the β-(1,4)-glycosidic bonds in peptidoglycan and thereby destroying many bacterial cells. Antibiotics such as penicillin commonly target bacterial cell wall formation (of which peptidoglycan is an important component) because animal cells do not have cell walls.

Structure

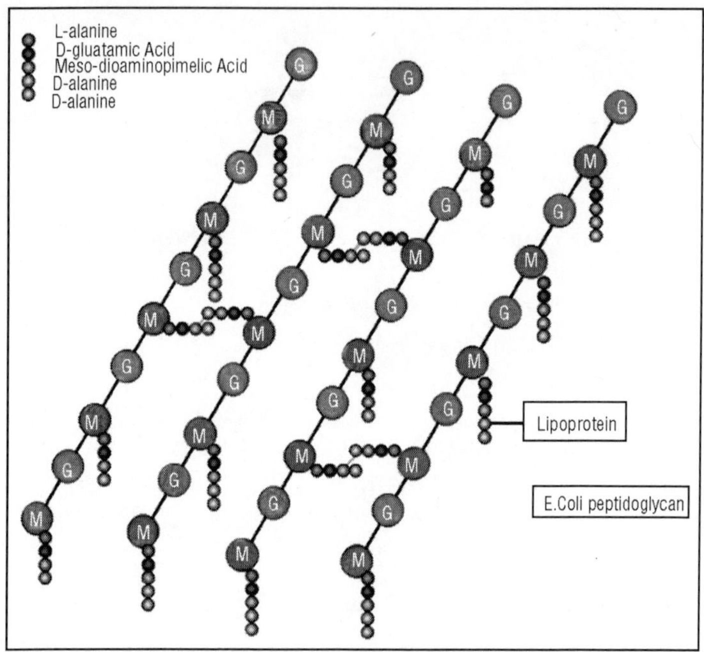

Fig. Structure of Peptidoglycan

The peptidoglycan layer in the bacterial cell wall is a crystal lattice structure formed from linear chains of two alternating amino sugars, namely *N*-acetylglucosamine (GlcNAc or NAG) and *N*-acetylmuramic acid (MurNAc or NAM). The alternating sugars are connected by a β-(1,4)-glycosidic bond. Each MurNAc is attached to a short (4- to 5-residue) amino acid chain, normally containing D-alanine, D-glutamic acid, and mesodiaminopimelic acid. These three amino acids do not occur in proteins and are thought to help protect against attacks by most peptidases. Cross-linking between amino acids in different linear amino sugar chains by an enzyme known as transpeptidase result in a 3-dimensional structure that is strong and rigid. The specific amino acid sequence and molecular structure vary with the bacterial species.

LIPOPOLYSACCHARIDES

LIPOPOLYSACCHARIDES

These complex compounds (LPS) are the endotoxic O-antigens found in the cell walls of Gram-negative bacteria (S-lipopolysaccharides) but are also present in one fungus, they may cause several pathophysiological symptoms, such as fever, diarrhea, blood pressure decrease, septic shock, and death. A lipid part (Lipid A) forms a complex with a core polysaccharide part through a glycosidic linkage. The core part is linked to a third external region of a highly immunogenic and variable O-chain polysaccharide or O-antigen made up of repeating oligosaccharide units. The latter region of the LPS molecule is responsible for bacterial serological strain specificity.

Lipid A consists of a backbone of β-1,6-glucosaminyl-glucosamine with two phosphoester groups in the 1-position of glucosamine I and in the 4-position of glucosamine II. The 3-position of glucosamine II forms the acid-labile glycosidic linkage to the long-chain polysaccharide. The other groups are substituted (in Escherichia) with hydroxylated fatty acids as hydroxymyristate (two ester-linked and two amide-linked) and normal fatty acids (laurate). The fatty acid profiles of lipopolysaccharides differ markedly from those of the phospholipids from the same species.

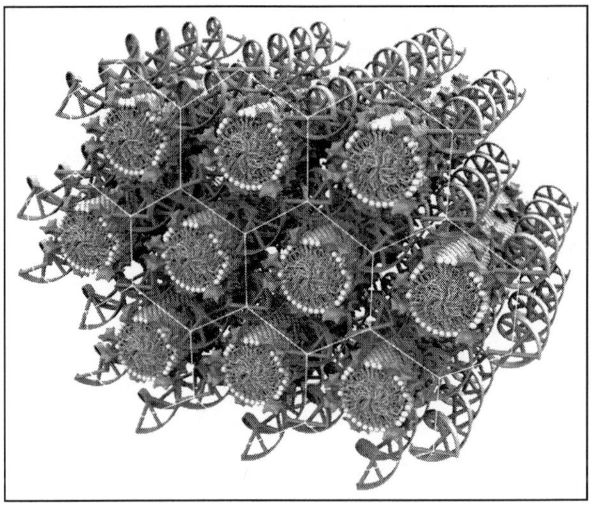

Fig. Structure of Lipid A of *Escherichia*

Although the hydroxylated fatty acids are generally 3-hydroxyalkanoic acids, 2-hydroxyalkanoic acids are also common. Exceptionally, some bacteria seem to lack hydroxy acids such as some species of Gardnerella, Thermus, Thiobacillus, Bacteroides, Sorangium, Achromobacter, and Brucella. In most organisms, the major hydroxy acids are straight-chain, even-carbon compounds (from 12 to 18 carbon atoms). Odd-carbon acids are minor components (13 to 17 carbon atoms) in some enterobacteria, some Vibrio, some Pseudomonads and Bordetella. They are also found in Pectinatus, Veillonella and Selenomonas.

Branched-chain hydroxy fatty acids are also present in Pseudomonas, Moraxella, Desulfovibrio, Capnocytophaga, Flavobacterium, Legionella species (3-OH i-11:0 up to 3-OH i-17:0). Few unsaturated hydroxy fatty acids have been described, but 2-OH-18:1 is present in some Pseudomonas and Azospirillum species. Lipopolysaccharides are the major toxins of Gram-negative bacteria (endotoxin). The only Gram-positive bacteria that possesses LPS is *Listeria monocytogenes*, an infective agent in milk. While the polysaccharide is non-toxic, the lipid A part is responsible for the toxic activities of these bacteria which result in septic shock.

CARBOHYDRATE METABOLISM IN FISH

Much of the carbohydrates that enter the diets of animals, including fish, is of plant origin. Carnivorous fish like the Atlantic salmon and the Japanese yellowtail, therefore, deal with little carbohydrate. Indeed, experiments have shown that these species are ill-equipped to handle significant quantities of raw carbohydrate, in their diets. On the other hand, omnivores such as the common carp and the channel catfish are able to digest fair amounts of carbohydrates in their diets. The grass carp, a herbivore, subsists primarily on a vegetarian diet.

DIGESTION, ABSORPTION AND STORAGE

The ability of animals to assimilate starch depends on their ability to elaborate amylase. All species of fish have been shown to secrete α -amylase. It has also been demonstrated that activity of this enzyme was greatest in herbivores. In carnivores such as the rainbow trout and sea perch, amylase is primarily of pancreatic origin whereas in herbivores the enzyme is widespread throughout the entire digestive tract. In Tilapia mossambica the pancreas has been shown to be the site of greatest amylase activity followed by the upper intestine. Although the digestion of starch and dextrin by the carnivorous rainbow trout was shown to decrease progr-essively as levels of the carbohydrates were increased beyond the 20 percent level, the fish could effectively utilize up to 60 percent glucose, sucrose or lactose in the diet. This demonstrates that, contrary to earlier belief, carnivorous fish are capable of efficiently utilizing simple carbohydrate as a primary energy source.

The crystalline structure of starch appears also to influence its attack by amylase as evidenced by the two-fold increase in metabolizable energy content of fully cooked (gelatinized) maize in feeding trials with channel catfish. Rainbow trout have also been shown to have a higher tolerance for carbohydrate (present as wheat starch) in the diet when it was cooked. The process of gelatinization involves both heat and water. If an aqueous suspension of starch is heated, the granules do not change in appearance until a certain critical temperature is reached. At this point some of the starch granules swell and simultaneously lose their crystallinity. The critical temperature is that at which hydrogen bonds of the starch molecule loosen to permit complete hydration, leading to a phenomenon known as "swelling".

Alpha-amylase, promotes a more or less random fragmentation of the starch molecule by hydrolyzing at the a -D-(1→ 4) glucosidic bonds in the inner and outer chains of the compound. The result of complete hydrolysis of the amylose component are maltose and D-glucose, while the amylopectin component is reduced to maltose, D-glucose and branched limit dextrins. As a consequence of these action patterns by α -amylase on starch, other enzymes are needed for complete hydrolysis of starch to D-glucose in fish. In this regard, it has been demonstrated that even the carnivorous sea bream possess the ability to digest maltose.

On the other hand, cellulase and α -galactosidase have not been shown to be secreted by fish although cellulase of bacterial origin is present in the gut of most species of carps. The lack of α -galactosidase may partly explain the poor response by fish to dietary soybean meal which contains significant levels of the galactosidic oligosaccharides raffinose, and stachyose.

As has been pointed out earlier, these oligosaccharides do undergo enzymatic hydrolysis during the germination process to yield galactose and sucrose. It would, therefore, appear that the nutritive value of soybean meal will be enhanced if the bulk of this indigestible starch is first transformed. This can be achieved by soaking the beans for 48 hours prior to processing for meal production. It should also be pointed out that the nutritive value of pulses and other legume seeds can likewise be improved for fish since oligosaccharides constitute a large portion of the carbohydrates in legume seeds.

Data on glucose absorption by fish are scanty. Work with goldfish has shown that active transport of glucose is coupled with Na^+ transport as in most mammals. It is generally believed that absorption takes place on the mucosal surface of intestinal cells. The mono-saccharides which result from carbohydrate digestion consist primarily of glucose, fructose, galactose, mannose, xylose and arabinose. Although the rates of absorption of these sugars have been determined for many land mammals, similar information for fish is not available.

Glucose does not appear to be a superior energy source for fish over protein or fat although digestible carbohydrates do spare protein for tissue building. Also, unlike in mammals, glycogen is not a significant storage depot of energy despite evidence of an active and reversible Emden - Meyerhoff pathway in fish. The more efficient metabolism of amino acids over glucose for energy could be due to the ability of fish to excrete nitrogenous waste as ammonia from their gills without the high cost of energy in converting the waste to urea.

OTHER FACTORS AFFECTING METABOLISM

Apart from genetic adaptation, climatic factors also play an important role in carbohydrate metabolism in fish. Acclimation in fish, in essence, reflects enzyme acclimation, since the animal's ability to survive depends largely upon its ability to carry out normal metabolic functions. Some enzymes for metabolic acclimation show good compensation while others do not. The enzymes associated with energy liberation (enzymes of glycolysis, pentose shunt, tricarboxylic acid cycle, electron transport and fatty acid oxidation) exhibit temperature compensation whereas, those enzymes dealing largely with the degradation of metabolic products show poor or reverse compensation.

Table. Enzymes Subject to Metabolic Acclimation
Enzymes Exhibiting.

Enzymes exhibiting compensation	Enzymes exhibiting reverse or no compensation
phosphofructokinase	catalase
aldolase	peroxidase
lactic dehydrogenase	acid phosphatase
6-phosphogluconate dehydrogenase	D-amino acid oxidase
succinic dehydrogenase	Mg-ATP ase
malic dehydrogenase	choline acetyl transferase
cytochrome oxidase	acetylcholine esterase
succinate-cytochrome C reductase	alkaline phosphatase
NAD-cytochrome C reductase	allantoinase
aminoacyl transferase	uricase
Na-K-ATPase	amylase
protease	lipase
	malic enzyme
	glucose-6-phosphate dehydrogenase

It is interesting to note that two key enzymes involved in carbohydrate metabolism, amylase and glucose-6-phosphate dehydrogenase, together with an enzyme involved in fat digestion, lipase, show no temperature compensation. It is not certain if this is in any way connected with the cessation of feeding by fish at low temperatures.

The molecular mechanism of thermal acclimation are not well understood and may consist of changes in synthesis or amounts of a given enzyme. Differences in kinetics, changes in the proportion of isoenzymes suitable for particular temperatures, and changes in co-factors such as lipids, co-enzymes, or other factors such as pH and ions may be important in the animal's adjustment to temperature changes.

ENERGY TRANSFORMATION

Despite species differences in the tolerance of dietary carbohydrates it is generally believed that the principal end-product of carbohydrate digestion, glucose, is metabolized in a manner prevailing in all cells, i.e., via the reversible Emden-Meyerhoff pathway. In this pathway, glucose has only one principal fate: phosphorylation to glucose-6-phosphate. The major metabolic transformations are depicted as follows:

All transformations proceed with a loss of free energy. Thus, the formation of two moles of lactate from glucose-6-phosphate occurs with free energy change of ΔG^o = -22000 cal/mole. The net result is the formation of four molecules of ATP. A functional reversal of this transformation can only occur via a different sequence requiring the input of six ATP molecules per mole of glucose-6-phosphate recovered.

Cells do not store glucose or glucose-6-phosphate. The readily available storage form is glycogen which is made from glucose-1-phosphate by one pathway and returned by another. Although in mammalian cells glucose-6-phosphate is transformed into fatty acids, such transformation does not appear to take place in fish. Studies with the common carp indicate that the precursor for lipogenesis is citrate formed when amino acids are actively metabolized through the tricarboxylic acid cycle.

The major form of utilizable energy in all cells is ATP. In most cells this energy currency is generated by the oxidation of NADH by the mitochondrial electron-transport systems. The reductants of NAD^+ for this process are intermediates derived from the TCA cycle and fatty acids.

TCA Cycle

The TCA cycle showing enzymes, substrates and products. The abbreviated enzymes are: IDH = isocitrate dehydrogenase and α-KGDH = α-ketoglutarate dehy-drogenase. The GTP generated during the succinate thiokinase (succinyl-CoA synthetase) reaction is equivalent to a mole of ATP by virtue of the presence of nucleoside diphosphokinase.

The 3 moles of NADH and 1 mole of $FADH_2$ generated during each round of the cycle feed into the oxidative phosphorylation pathway. Each mole of NADH leads to 3 moles of ATP and each mole of $FADH_2$ leads to 2 moles of ATP. Therefore, for each mole of pyruvate which enters the TCA cycle, 12 moles of ATP can be generated.

CITRATE SYNTHASE (CONDENSING ENZYME)

The first reaction of the cycle is condensation of the methyl carbon of acetyl-CoA with the keto carbon (C-2) of oxaloacetate (OAA) to form citrate. The standard free energy of the reaction, -8.0 kcal/mol, drives it strongly in the forward direction. Since the formation of OAA from its precursor is thermodynamically unfavorable, the highly exergonic nature of the *citrate synthase* reaction is of central importance in keeping the entire cycle going in the forward direction, since it *drives* oxaloacetate formation by mass action principals.

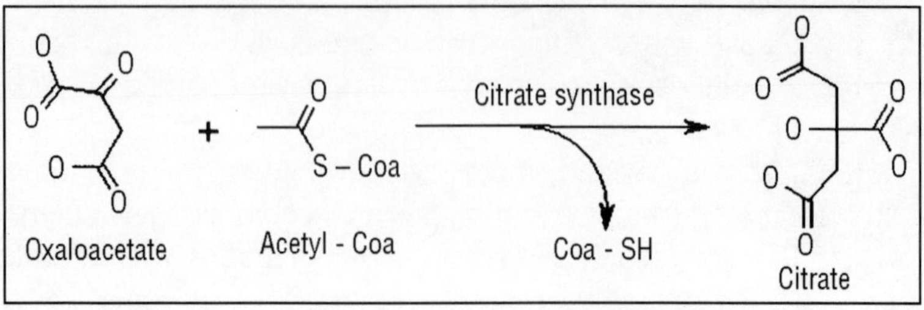

Citrate synthase
Oxaloacetate + Acetyl - Coa → Citrate + Coa - SH

When the cellular energy charge increases the rate of flux through the TCA cycle will decline leading to a build-up of citrate. Excess citrate is used to transport acetyl-CoA carbons from the mitochondrion to the cytoplasm where they can be used for fatty acid and cholesterol biosynthesis.

Additionally, the increased levels of citrate in the cytoplasm activate the key regulatory enzyme of fatty acid biosynthesis, *acetyl-CoA carboxylase* (ACC) and inhibit PFK-1. In non-hepatic tissues citrate is also required for ketone body synthesis.

6

Consumption of Seafood Processing Industries

INTRODUCTION

One of the most important challenges facing the country is providing remunerative prices to farmers for their produce without incurring the additional burden of subsidies. This challenge could be addressed if the processing level and value addition of the agricultural raw produce can be enhanced to meet the growing demand for processed foods.Seafood processing has an important role to play in linking Indian agriculture to consumers in the domestic and international markets.

The role of food processing becomes critical since food production is targeted to double in the next 10 years. With low farmer price realizations and significant wastage in the food supply chain even with the current level of production, only processing of food can secure farmer incomes against a slump in prices as well as reduce wastages. Further, a vibrant food processing industry can be a catalyst for crop diversification. The impact of increased economic growth in agribusiness through food processing can play a significant role in reducing rural poverty and increasing rural incomes.Seafood processing leads to significant employment generation- not only directly but also indirectly across the supply chain in production and procurement of raw materials and distribution of food products to consumers.

CONDITIONS AND TRENDS OF CONSUMPTION

In feeding stuffs very great quantities of fats disappear but, with a few unimportant exceptions, they always do so as a constituent of some element of the feed and not as such. Thus the amounts of fat that disappear with grain and legumes fed to animals are truly enormous. In food, as in feeding stuffs, very large quantities of fats are consumed incidentally in the ingestion of meats, fish, poultry, milk, and other animal products and in the form of fats and oils naturally contained in cereals, legumes, fruits, vegetables, and other plant products. However, in the human dietary fats and oils are also consumed practically as such, for example as butter, salad oils, and cooking fats.

QUANTITATIVE DATA UNSATISFACTORY

Not only because fats and oils disappear both in the arts and in the dietary in a vast variety of ways, but for other important reasons, the quantitative study of the consumption of fats and oils encounters great difficulties. In the first place, the very concept of consumption is elusive. Presumably one should exclude, in spite of its potential significance, the fat content of large numbers of meat animals that perish from natural causes and from which no attempt is made to recover the fats.

Commercial consumption is a significant concept, but it excludes a large amount of fats in products that do not enter into trade, and a large amount that are incidental components of meats and other products that do enter into trade, for example, tankage and garbage. Wastes in the process of recovery are considerable; hence the fat content of the product treated may be considerably greater than the fats recovered as such plus those incidentally retained. The amounts ingested by animals and by human beings would be worth knowing, but they would exclude large quantities that are wasted in various ways on the farms and in households and public eating places. Furthermore, a distinction may be drawn between intermediate and final consumption. Large amounts of fats are contained in cereals, nuts, milk, oil cake, and other products fed to animals; these are in part used up by the animals, and in part are converted or stored up and appear in dairy products or slaughtered animals. Hence a total of the fats consumed by animals and those derived from animals would contain, in effect, considerable duplication. The difficulties are enhanced by the fact that our statistical information is defective especially in respect to the data on animal fats.

For the United States, the number of animals killed in inspected slaughterhouses is known; the total slaughter at wholesale is reported by the census of manufactures; but only a guess can be made at the number of animals in other uninspected, principally rural, slaughter. Moreover, there is no clear-cut idea of the fat content per average animal slaughtered; the long-time trend seems to be downward, because younger and lighter animals are coming to slaughter in response to the increasing demand for lamb instead of for mutton, for bacon-type hogs instead of for the lard type, and for baby beef instead of for three- to four-year-old steers. To take even the best ratios derived from packing-house practice and apply them to the total wholesale slaughter of cattle, calves, sheep, and hogs as reported in the biennial censuses, and also to the farm and retail slaughter for which the data are officially admitted to be crude estimates, would be little better than intelligent guessing. At the best, there are gaps in the information so large as to make statistical conclusions on several important components, and on fats and oils as a whole, exceedingly unreliable. It would be of interest to know, for each of the several sources of fats:

- The amount wasted for lack of any attempt at recovery;
- The amount ingested;
- The amount wasted in households and public eating places;
- The amount fed to animals; and
- The amount used in industrial production other than food.

We should like these data for fats produced, fats imported, and fats exported. We should like to distinguish between gross and net consumption. Unfortunately, for many of the sources of fats the statistical information is exceedingly defective and it is difficult to reach reliable bases upon which to make the estimates necessary for arriving at supplementary data. We have made attempts to reach a rough approximation to the amounts of fats and oils available for human consumption or industrial use in the United States in recent years, but at several points gaps are so wide and the procedure is so open to objection that we do not feel justified in presenting here even a preliminary approximation, lest in spite of reservations and qualifications it should be taken for more than it would be worth.

In the discussion that follows, therefore, we have incorporated quantitative data only to a very limited extent. The two principal items omitted, fat of dressed meats and milk fats (including butter), are so difficult to estimate that we have not included them; yet their sum probably exceeds the sum of the fats and oils here itemized. They go largely into edible uses, except as milk is fed to animals, though in considerable measure meat fats are wasted in the household or subsequently recovered from household wastes for industrial use. The principal other items omitted are fats incidentally consumed in fish, poultry, grains, vegetables, and fruits. Judging from estimates of Raymond Pearl for the period, the sum of these items is probably less than a billion pounds a year, exclusive of the fat in grain fed to cattle.

FATS AND OILS IN DIET

The function of fats and oils in the diet is mainly to furnish energy to operate the animal machine. In the body, fats are burned as truly as though they were burned in a candle or under a steam boiler, and the end-products of the combustion are the same—carbon dioxide and water. Moreover, the amount of energy they furnish is the same, namely from 9 to 9.4 calories to the gram, whether they are burned within or outside an animal body. (A large calorie, which is the one here used, is the quantity of heat necessary to warm 1 kilogram of water from 0 deg Centigrade to 1 deg Centigrade. A small calorie is the quantity of heat necessary to warm 1 gram of water from 0 deg Centigrade to 1 deg Centigrade.) They yield more energy than the other important classes of foodstuffs, such as carbohydrates and protein (albumin) Conditions and Trends of Production—

Vegetable oils major products), which furnish from 3.8 to 4.2 calories to the gram. Neither fats nor carbohydrates ingested are wholly burned at once unless they are needed for the operation of the animal machine.

If not so needed, carbohydrates are for the most part completely changed in character by conversion into the fat characteristic of the species. This is stored in the body as a reserve against the possibility of a future period of food shortage. Indeed, most of the fat of domesticated animals is produced from starch. Thus the lard of the hogs of the corn belt is mainly derived from the starch of corn (maize).

It has the characteristics of fat normal to the animal, is normally stiff, and hence is especially prized. Unlike ingested carbohydrate, the fat of the food, if it is not at once burned, is changed comparatively little. For the most part it is deposited in but slightly modified form with other fat, made by the animal from carbohydrate or protein, in the storehouses for fat—the adipose tissue under the skin, in and about the viscera, and elsewhere. Therefore, if the food fat differs in its properties from the fat natural to the animal and if there is a great deal of it in the feed, the storage fat formed by the deposition of food fat gives an abnormal character to the fat which is obtained from the animal after slaughter. The dietary fats may be divided into four groups:

- Those consumed as such on the table;
- Those employed in the preparation of foods;
- Those consumed incidentally in the ingestion of meats and other animal products; and
- Those consumed incidentally in cereals, legumes, fruits, and vegetables.

The fats consumed as such on the table in the United States are principally butter, butter substitutes (chiefly coconut and oleo oil), and salad oils (principally cottonseed, corn, and olive oils). The fats and oils used in the preparation of food are butter, lard, lard compounds (for the most part cottonseed oil), cottonseed oil as such, corn oil, peanut oil, butter substitutes, and, in certain types of confectionery, coconut oil and cacao butter. Fats and oils consumed incidentally in the ingestion of meats and dairy products are found principally in milk, cheese, poultry, eggs, beef, mutton and lamb, pork, and fish.

The fat of meats is of course to a material extent not ingested but wasted except in so far as it is recovered in garbage and similar greases. The fats of the cereals, legumes, fruits, and vegetables as a class are important in the aggregate nutritionally, but have little direct commercial importance, since few foodstuffs are purchased on the basis of fat content. Practically all foodstuffs except sugar, water, and certain condiments contain some fat. Fresh fruits and vegetables contain merely traces; nuts, on the other hand, are rich in fats, as are also chocolate products. Oat meal and corn meal that is made without removing the germ are relatively rich in fat; wheat flour is poorer.

A study of the trends of consumption encounters the difficulty that statistics of fat in the diet over a series of decades are not available. Certain inferences seem nevertheless warranted. Trends of consumption may be grouped in two classes according as they are due primarily to changes in habits or primarily to substitutions made by manufacturers of which the consumer may or may not be aware. The two classes of course overlap, for in certain cases both factors play a part. Thus a manufacturer may create a new product which leads to a new food habit or, vice versa, consumer demand may lead to substitution or creation of a new product by the manufacturer.

INFLUENCE OF CHANGES IN FOOD HABITS

Change of food habits may be of two sorts. It may be purely quantitative or it may be qualitative. By a quantitative change is meant increase or decrease in consumption of a given fat rather than the substitution in the diet of one fat for another. By qualitative change is meant primarily a substitution of one fat for another. However, since for physiological reasons the intake of food is practically constant for any individual under any given set of circumstances, a decrease in fat ingestion usually involves either a substitution for it of some other kind of food, or of some other kind of fat. Conversely, an increase usually involves a decrease in consumption of some other kind of food or of fat. Since all fats and oils have very nearly the same food value, one can be substituted for another without change in the volume of food ingested. This statement applies to energy values. Certain fats contain small quantities of chemical substances of unknown chemical nature but of great importance to health. These are known as vitamins or food accessories.

Different fats contain different amounts of them depending upon their origin, method of preparation, and other factors. From the point of view of their vitamin content all fats are not of equal nutritive value; but these are considerations. However, if another food be substituted, the volume of the diet is thereby necessarily increased since no other food has, weight for weight, so great an energy value. It follows that if the replacement of fat in the diet by other foods goes too far the diet must become very bulky if it is still to furnish the same energy value, and this fact sets mechanical limits to the substitution of other foodstuffs for fat in the diet.

Since the caloric requirements of an individual, as stated above, are constant under given conditions, it follows that under these circumstances a reduction in consumption of a fat involves the substitution for it of either an equal amount of some other fat or a corresponding amount of some other foodstuffs. But conditions do not remain constant for any person. He grows old and requires less food. He may become stouter or leaner, then consuming either more or less food. He may change his means of livelihood from one requiring hard manual labour involving high food requirements to a sedentary occupation requiring a low food intake. He may migrate from a very cold climate where food requirements are relatively high to a hot one where they are relatively low. In any nation all these changes are taking place in some individuals in one direction, in others in the opposite direction. For short periods, the result is a reasonably constant consumption.

Over a long period of time this is no longer true. The age distribution of populations changes. If the birth-rate falls and the average span of life lengthens, the proportion of old people who consume little food increases, the proportion of young people who consume much decreases. If it becomes unfashionable to be stout, less food is consumed by the nation as a whole. If the proportion of manual labourers becomes less and that of machine tenders and sedentary workers greater, the per capita consumption of food tends to fall. Since fat is the most concentrated form of food energy, its intake tends to be reduced rather more than that of other forms of food.

In the United States exactly these changes have been taking place for decades; and there is evidence of a corresponding decline in the nutritional use of fats and oils, as part of a general reduction in the per capita food requirements. This reduction is the result of substitution of machine labour for man labour, the decline of outdoor employment in severe winter weather, the improved heating of buildings, the trend to lower average body weights of the people generally, and changes in the age distribution of the population. The population is acquiring more sedentary characteristics, needs less food, and the average per capita requirements of foodstuffs in terms of calories are automatically reduced. Most of these factors are the result of the economic evolution of the country and in essence reflect improvement in the standard of living.

The rise of the standard of living is responsible for certain other changes in food habits. It results in a diversification of the diet with increase in the use of dairy products, fruit, vegetables, and sugar, and decline in the ingestion of cereals and fats. This general statement does not hold for the fat of milk, the use of which is on the increase. Fat consumed incidentally to the ingestion of meats and dairy products has always been heavy. The consumption of milk and poultry fat is increasing, that of beef, sheep, and hog fats is declining. Plain cooking is being replaced by more fancy cooking, which includes more discriminating culinary uses of fats and oils. American practice has departed from the British custom of boiling vegetables without fat. Much of the crackers and bakers' bread consumed now contains shortening.

These facts might seem to suggest increased use of fats and oils. A survey of the entire field, however, suggests that the net result is a lower per capita ingestion of fats in general. The heavy fat rations of hard workers are a thing of the past—fat-backs and sow-belly are no longer staples. We use more butter, but less hog fat. The current public taste for younger and lighter-weight animals represents a substantial reduction in fat. It seems fair to infer that, milk fat and poultry fat aside, the decline in the ingestion of fats applies to animal fats. Over a generation, indeed, it is possible that there may have been an absolute increase in the per capita consumption of vegetable oils with an absolute decline in the per capita consumption of animal fats outside of milk fat.

INFLUENCE OF CHANGES IN FOOD MANUFACTURE

The changes in trends of food-fat consumption are not all by any means due solely to change in habits. Some of those that are quantitatively extremely important are due to substitutions on the initiative of manufacturers of food-fat preparations—for example, the substitution of lard compounds for lard—or to the increasing availability of new types of fat—for example, coconut oil. Some of the possibilities of substitution in food uses have already been touched upon incidentally.

However, since dietary consumption is the major and the higher or premium use of fats, the practice and possibility of substitution require elabouration because they are basic to any consideration of the commodity economics of this important group of raw materials. The substitution of one fat for another, indeed the mere possibility of such substitution, naturally has the greatest influence upon prices.

Given sufficient price inducement, one fat may in some cases displace another with the widest repercussions upon the producer, the farmer, the trader, and the manufacturer. It is one of the purposes of the *Fat and Oil Studies of the Food Research Institute* to present studies from time to time upon far-reaching movements of this general character. That considerations of price, costs of conversion and refining, lack of technological skill, and legislation in the interests of public health and sanitation limit the volume of inedible fats turned to edible uses, has already been pointed out.

But there are other factors as well that limit not merely the diversion of inedible fats to edible uses but also the substitution of one edible fat for another. The most prominent of these are greater or lesser adaptability of different fats to use in the preparation of different foodstuffs, and psychological and sentimental factors that in the food industries play a greater part than in the arts.

CONSUMPTION OF FOOD PRODUCTS

The consumption of food in India was estimated at approximately INR 4,000 bn at 1993-94 prices (INR 7,250 bn at 2001-02 prices) in the year 2001 -02, growing at a CAGR of 7.8per cent (1996-2002).Seafood consumption has been driven primarily by increase in per capita expenditure on food items. In the period 1996-2002, the per capita expenditure grew by 6 per cent per annum and population increased at the rate of 1.6per cent per annum. Urban food consumption has grown at 9.1 per cent annually and is estimated at INR 1,540 bn (1993-94 prices). Rural food consumption, on the other hand, has grown at

7per cent annually and is estimated at INR 2,550 bn (1993-94 prices). Uttar Pradesh is the largest consumer market for food products followed by Maharashtra, West Bengal, Bihar and Andhra Pradesh. Among the fastest growing markets are the North East states, Gujarat and Andhra Pradesh.

The share of grain-based products is the highest in the Indian consumer's food basket, followed by milk and milk products, vegetables, edible oil and meat products. The growth rates for fruit, vegetables, meat and dairy products is higher than grains and pulses, indicating a shift in consumption pattern. This also implies the need for diversification in agricultural production to match the change in consumption patterns.

CONSUMPTION OF PROCESSED FOOD PRODUCTS

The market size for processed foods for year 2003-04, at current prices is estimated at INR 4,600 bn, which is about 53 per cent of total food consumption. This includes processing in unorganized sector in dairy (Halwais) and grains (chakkis).The market size excluding these segments is about INR 3,300 bn. The market size of processed foods at factory cost for 2003-04 is estimated at INR 2,100 bn[3] implying about 60 per cent mark-up from factory cost to market price. However, primary processed products constitute as much as 62 per cent of processed foods consumed, with value-added products being the balance 38per cent.

'excluding consumption of alcoholic beverages and out-ofhome consumption

"excluding out-of-home consumption but including alcoholic beverages and processing in unorganized sectors in dairy (halwais) and grain milling (chakkis)

The low share of value added products in food consumption is due to low level of processing across agricultural products. The key impediments to growth of processed foods include demand-side, supply-side and regulatory issues. The demand side factors include:

- Low per capita income
- Socio-cultural factors such as preference for freshly cooked products as compared to packaged products and availability of low-cost domestic help

The supply side factors constraining the demand for processed foods include:

- Long and fragmented supply chain with independent players engaged in various value-addition activities such as input supplies, cultivation, processing, distribution, storage and retailing (unlike in other major agriculture/food producing countries, where a large number of food companies are integrated across the chain. *e.g.* ADM, Cargill, Tyson Foods)
- High cost of raw material (due to low productivity and poor agronomic practices) and lack of varieties amenable for processing

- Lack of scale
- High cost of packaging
- Lack of infrastructure (cost of power, lack of cold chains, storage, handling and transportation)
- High cost and poor quality of distribution.

CHEMICALS IN FOOD PROCESSING

The same over commitment to potentially risky methods is found in other phases of food production. Complex problems of storage, cleansing, handling, refining, cooking, mixing, heating and packaging are often solved by the use of chemical additives. A half century ago the yellowish tint in freshly milled flour was removed slowly by aging; today it is removed rapidly by oxidizing agents. The miller no longer has to store flour and wait for it to mature. Emulsifiers are employed, partly to "improve" the texture of a product, partly to speed up processing. Their "functional advantages" in ice cream, for example, are that the "mix can be whipped faster, the product from the freezer is dryer and the mix has less tendency to 'churn' in the freezer."

Chemical antioxidants are used to help prevent oils and fats from becoming rancid, a problem that becomes especially pronounced when natural antioxidants are removed during large-scale processing. Chemical preservatives are added to food to permit longer storage in wholesale depots and to extend the "shelf life" of a product in retail outlets. In time chemicals are turned from mere adjuncts of food production into "technological necessities." Their use may result in new machines and facilities, the abandonment of old processing methods and a broad reorientation of technology to meet the new chemical requirements.

The fact that a chemical is technologically useful does not mean that it is biologically desirable. The two phrases are occasionally used interchangeably, as though they were synonymous. Some of the chemicals most widely used in the production of food have been found to be harmful to animals in labouratory tests. Nitrogen trichloride, or agene, was employed for decades as a bleaching and maturing agent in flour before it was discovered, in 1946, that the compound produces "running fits," or hysteria, in dogs.

The use of agene in the bread industry has since been discontinued. Although there is no evidence that agenized flour causes nerve disorders in man, no one can say with certainty that cereal products treated with agene are completely harmless to consumers, especially if eaten day after day for twenty years or more. The toxicant in agenized flour "produces nervous disturbances in monkeys, but not epilepsy in man," observed the late Anton Carlson, one of the most distinguished physiologists of his day. "That is not sufficient for me, because there are many other types of nervous disturbances that may be aggravated or introduced by this chemical."

Another group of chemicals, notably certain types of polyoxyethylene derivatives, were widely used in the United States as emulsifiers before they were banned as potential hazards to consumers. Former F.D.A. Commissioner Charles W. Crawford is reported to have described the agents as "good paint removers." According to data furnished to the Delaney Committee, rats that had been fed two commercial preparations of polyoxyethylene-lb/> derived emulsifiers showed blood in their feces and developed a variety of kidney, abdominal, cardiac and liver abnormalities.

These preparations were used primarily as bread softeners but also found their way into other foods. The polyoxyethylene derivatives constitute a highly suspect family of emulsifying agents, yet some of them are still added to cake mixes, cake icing, frozen desserts, ice cream, candies, dill pickles, fruit and vegetable coatings, vitamin preparations and many food-packaging materials. Similar examples can be culled from other groups of chemical aids. In general, the toxic effects of many commonly used compounds are difficult to determine, but individually and in groups they have aroused suspicion and concern. "The fortification of oils and fats against oxidation and rancidity is another field in which there are serious questions of possible toxicity," warns F. J. Schlink, technical director of Consumers' Research and editor of *Consumer Bulletin*. "The natural antioxidants which delay rancidity are lost during the factory processing of refined oils and fats; then the attempt is made to restore the antioxidation qualities by addition of fat-stabilizing substances. Most of the materials that have antioxidation properties are known to be toxic to some degree."

The term "technological aid" has been used loosely to include food additives that could be dispensed with entirely if a manufacturer improved his processing methods and the quality of his product. The question of whether a chemical, instead of being legitimately employed to furnish distant consumers with a perishable product, was being used rather as a device for masking inferior ingredients and unsanitary methods arose in sharp form early in the century, when sodium benzoate was employed extensively as a preservative in catsup. Harvey Wiley, the first administrator of the federal food and drug law, regarded sodium benzoate as a hazard to consumers.

Supported by data from feeding experiments performed with human beings, Wiley claimed that the preservative causes injury to the digestive system. He was convinced that catsup need not deteriorate readily if clean facilities and proper ingredients are used in preparing it. Challenged on this point by Charles F. Loudon, a canner in Terre Haute, Indiana, Wiley appointed a bacteriologist, Arvill W. Bitting, to work at the Loudon factory and prepare a formula for a stable catsup preparation without artificial preservatives. Bitting was successful. Moreover, after inspecting numerous canning factories and examining many brands of catsup, Bitting concluded that manufacturers who needed sodium benzoate often used inferior products and were careless about sanitation. "The whole spirit and tradition of the pure-food law is against the use of preservatives and substitutes," declared Mrs. Harvey Wiley more than forty years after the Loudon episode. "The intent of the law is to encourage the manufacture of high-grade ingredients, not to try to hide inferiority and cheapness by the use of chemicals and preservatives." If this is so, the spirit of the law has been ignored for decades.

The most up-to-date surveys of chemical additives in foods show that preservatives are added to cheese, margarine, cereal products, jams, jelly and many other processed foods. Uncooked poultry is dipped in solutions of antibiotics. Aside from the damage to public health that may arise from the use of questionable chemicals as preservatives, the longer "shelf life" acquired by some of these foods may cause a loss of valuable nutrients. Many chemical additives in foods perform no technological or nutritional function whatever. They cannot be regarded as materially useful by any stretch of the imagination. These chemicals enhance the colour of a food or make certain products feel soft and newly prepared; in some instances they have been used to replace costly but valuable nutrients with inferior ones. In the least objectionable cases, an additive will deceive the consumer without impairing his health.

The application of an innocuous vegetable dye to some foods often leads the consumer to believe that he is acquiring a better, more wholesome, or tastier product than is actually the case. In the most objectionable cases, a chemical is toxic to a greater or lesser degree. This type of additive not only serves to change the appearance or conceal the nutritive deficiency of a food; it exposes the consumer to a certain amount of damage.

Nitrates and nitrites, for example, are used to impart a pink colour to certain brands of processed meat. Ostensibly, this is their sole function. The addition of these chemicals to meat, especially frankfurters and hamburger meat, would be objectionable if only because of the deception which makes it difficult for the housewife to distinguish between high-quality and low-quality products. But this is not the only deception perpetrated on the consumer. Used together with salt, nitrite compounds definitely extend the "shelf life" of processed meat products.

When nitrites are ingested, they react with hemoglobin in the blood to form methemoglobin and, like carbon monoxide, reduce the hemoglobin's capacity to carry oxygen. An individual who consumes three to four ounces of processed meat containing 200 parts per million of sodium nitrite (a permissible residue) ingests enough of the compound to convert from 1.4 to 5.7 per cent of his hemoglobin to methemoglobin. Ordinarily, this percentage is insignificant. But if the same individual is a heavy smoker and lives in an air-polluted community, the cumulative loss in oxygen-carrying capacity produced by the nitrites in the food and by the carbon monoxide in tobacco smoke and motorcar exhaust can no longer be dismissed as trivial.

Sodium nitrite is highly toxic in relatively small amounts. About four grams of the compound constitutes a lethal dose for adults. Although Lehman regards 200 parts per million of sodium nitrite as a safe residue, he notes that "only a small margin of safety exists between the amount of nitrite that is safe and that which may be dangerous. The margin of safety is even more reduced when the smaller blood volume and the corresponding smaller quantity of hemoglobin in children is taken into account. This has been emphasized in the recent cases of nitrite poisoning in children who consumed weiners and bologna containing nitrite greatly in excess of the 200 ppm permitted by Federal regulations. The application of nitrite to other foods is not to be encouraged."

Emulsifiers have been used in bread not only as softening agents, which can give stale bread the texture of newly baked bread, but as substitutes for nourishing ingredients. "The record of the bread-standards hearings contain evidence of distribution among bakers of advertising material advocating the replacement of fats, oils, eggs and milk by emulsifiers," George L. Prichard, of the Department of Agriculture, told the Delaney Committee. "The use of such products as components of food may work to the disadvantage of our farm economy by displacing farm products normally used. The record indicates that natural food constituents, such as fats and oils, probably will be reduced in many commercial bakery products if bakers are allowed to employ these emulsifiers."

This substitution affects more than 117 the 117 farm 117 economy, however; it also works to the disadvantage of the consumer. To illustrate this, the Delaney Committee report compared the ingredients in two cake batters prepared by the same company eleven years apart; during the interval emulsifiers were added to the product. The first batter, made in 1939, did not contain synthetic emulsifiers; the second, prepared in 1949, did. "On a percentage basis, the cake batter in 1939 contained 13 per cent eggs and 8.6 per cent shortening.

In 1949, the cake batter contained 6.3 per cent eggs and 4.8 per cent shortening, with somewhat less than 0.3 per cent of synthetic emulsifier." The report noted that a "synthetic yellow dye could be added to provide the colour formerly obtainable through the use of eggs. The utilization of synthetic yellow dye in commercial cake was practiced before the war. There are indications that the use of artificial colouring matter is increased when quantities of whole eggs or egg yolks are reduced in commercial cake formulas." To heighten the insult, among the most commonly used emulsifiers in 1949 were the polyoxyethylene monostearates. The yellow dye referred to in the Delaney Committee report was probably FDC Yellow No. 3(Yellow AB), which, until fairly recently, was added to many yellow cake mixes. The dye often contained impurities of a potent carcinogen and its use in foods was forbidden in 1959. For a number of years, however, both the emulsifier and the dye undoubtedly appeared in many brands of cake and reached large numbers of unsuspecting consumers.

Problems of this kind are not likely to disappear unless there is a basic change in the viewpoint of the F.D.A. "Inherited from the Wiley era is a too common misconception that all 'chemicals' are harmful and the related idea that any amount of a 'poison' is harmful," the F.D.A. observes in a brochure on food additives. "The fact is, of course, that chemical additives, or food *additives* as they are now being called, have brought about great improvements in the American food supply. Additives like potassium iodide in salt and vitamins in enriched food products are making an important contribution to the health of our people and yet it is a fact that both iodine and some of the vitamins would be harmful if consumed in excessive amounts. Many similar examples could be given to refute these common misconceptions."

This argument is grossly misleading. Iodine and vitamins are indispensable to human life, but coal-tar dyes and benzoic acid, for example, are not. If we adhered to a well-balanced diet of properly prepared natural foods, iodine and vitamins would never enter the body in toxic amounts. Coal-tar dyes, on the other hand, are suspected of being harmful to man in nearly any amount if consumed repeatedly and the kindest statement that can be made for the presence of benzoic acid in food is that the compound is "safe under the conditions of its intended use."

The formula "safe under the conditions of its intended use" exposes the consumer to serious risks when it opens the door to the use of food additives whose biochemical activity is not understood. Many unexpected problems may arise when such additives appear in food. A particular additive may seem to be relatively harmless to the organs of the body, but it may be carcinogenic on the cellular level of life. Another additive may produce insignificant or controllable effects when studied in isolation; brought into combination with various chemicals in food, however, it may give rise to toxic compounds.

The body, in turn, may make a toxic additive more poisonous in the course of changing it during metabolism. "At one time it was generally believed that whenever a toxic substance was absorbed the body was capable of calling upon special mechanisms for detoxifying the toxicant," Lehman observes. "It was believed also that the metabolic pathway that the toxicant followed always proceeded in the direction of the formation of a less toxic compound. Later work on the metabolism of drugs and toxic substances showed that special mechanisms did not exist, but that the toxicant was subject to the same metabolic influences as those which normally operate in the body.

The assumption that the metabolic product was less poisonous than the parent substance from which it was derived is also unwarranted simply because in many instances the toxicity of either the original substance and its conversion product is unknown. In other instances the metabolic product is even more poisonous than its parent. The conversion product, heptachlorepoxide, which is two or three times more poisonous than the parent substance, heptachlor [a widely used insecticide], may be cited as an example of this." Finally,seafood additives may cause allergic reactions that are likely to go unnoticed for many years. The reactions need not be severe to be harmful. Otto Saphir and his colleagues at the Michael Reese Hospital in Chicago have recently suggested that the development of arteriosclerosis may be promoted by the sensitivity of the body to allergenic compounds, notably certain antibiotics. By using sulfa drugs to produce allergies in forty-two rabbits, the researchers were able in eight months to cause degenerative changes in the arteries of thirty-one of the animals.

These changes closely resembled arteriosclerosis in man. According to a press account of Saphir's report, the rabbits' reactions "were not apparent on the surface." It would be very imprudent to assume that such effects are produced only by chemicals that cause severe or noticeable allergies. If the data of Saphir and his colleagues are applicable to man, additives with even minor allergenic properties cannot be dismissed as harmless. Today more than 3,000 chemicals are used in the production and distribution of commercially prepared food. At least 1,288 are purposely added as preservatives, buffers, emulsifiers, neutralizing agents, sequestrants, stabilizers, anticaking agents, flavouring agents and colouring agents, while from 25 to 30 consist of nutritional supplements, such as potassium iodide and vitamins. The remaining chemicals are "indirect additives"—substances, such as detergents and germicides, that get into food by way of packaging materials and processing equipment. Many chemical additives are natural ingredients, but a large number are not. The artificial substances that appear in food range from simple inorganic chemicals to exotic compounds whose biochemical activity is still largely unknown.

DISTRIBUTION OF FOODS IN INDIA

Food products are sold through over 5 million food and grocery stores in India. The organised food retail market is estimated at INR 20 bn (approximately 0.2 per cent of food expenditure in India).This is in contrast to developed countries, where food distribution is highly consolidated. The majority of food and food products are retailed through neighbourhood kirana stores. The kirana stores focus on dry food products in the absence of infrastructure for cold storage. Bulk of fresh produce is sold by vendors with push carts. Meat, poultry and marine products are primarily sold by small retailers in wet markets. Such produce is associated with low product quality, lack of variety and low hygiene levels.

Internationally,seafood retailers have played an important role in improving supply chain efficiencies such as developing storage and transportation infrastructure, training supply chain members on food hygiene and standards and providing scientific know how to farmers. The key impediments to growth of organised food retailing in India include lack of infrastructure, technology and capital.

Foreign Direct Investment (FDI) is not allowed in retailing with the exception of cash and carry formats. This restriction is based on the premise that entry of large international players will displace existing employment and reduce bargaining power of farmers. However, there is empirical evidence from other countries (USA, China) which highlights that opening of the retail sector to FDI has enabled employment creation, disintermediation, increased farmer access to information and wider choice and cost savings for the consumer.

EXPORTS OF AGRICULTURAL AND FOOD PRODUCTS

India has 1.5 per cent (INR 360 bn in 2003-04) share of global agricultural exports (approx USD 522 bn), despite its production leadership in agriculture. India's exports primarily constitutes commodity and primary processed items, where price realisations are low. In addition, many products are showing single digit or negative growth. The reasons for India's insignificant share in global trade include supply side factors such as lack of consistency in supply and quality, lack of cost competitiveness, and demand side factors such as non-tariff barriers, short product life cycles and perception of Indian food products in overseas markets.

Most exporters from India lack scale- for example the largest fresh produce exporter records annual sales of about INR 500mn.This has resulted in lack of economies in operations and renders them uncompetitive. Hence, exporters are not able to establish themselves as long-term players in the export market, rely on opportunistic businesses and are consequently, unable to develop technical and managerial expertise. These factors cumulatively lead to low investments by exporters in brand building, quality improvement and brand development.

STATUS OF FOOD DISTRIBUTION IN INDIA

Food accounts for the largest share of consumer spending. Seafood and food products account for about 53per cent of the value of final private consumption estimated at INR 8600 bn (2003-04 at current prices). Seafood products are sold through over 5 million food and grocery stores in India. The size of the organised food retail segment in India is estimated at INR 20 bn (approximately 0.2 per cent of food expenditure).This is in contrast to most of the developed countries, where food distribution is highly consolidated.

In countries such as Australia, France, Germany, Denmark and The Netherlands, the share of top the 10 retailers is more than 80per cent (USA ~ 35per cent, China ~ 15per cent). Most organised retailers in India are regional and use single formats such as convenience stores, supermarkets, hypermarkets or cash and carry. In contrast, most international retailers have multiple format models. Besides reflecting the stage of evolution of retailing in India, this also highlights the capital constraints of existing players in India, restricting their ability to make large investments. In India,seafood companies are much larger in size than the organised food retailers in India. In contrast, the retailers are larger in scale, than food companies, in developed countries. Given the channel power of retailers in these markets,seafood companies have started establishing partnerships with them to develop and test new products, share consumer information and undertake consumer promotions.

KEY DRIVERS

Increasing need for convenience: The Indian consumer visits about eight to ten outlets to purchase various food products which constitute the daily consumption basket. These outlets include neighbourhood kirana stores, bakeries, fruit and vegetable outlets, dairy booths and chakkies (small flour mills). With changing lifestyles, there is growing paucity of time, and convenience in food shopping is emerging as an important driver of growth of one-stop retail formats. Availability of quality retail space: Until the late 1990s, the high cost of real estate meant that organised food retail business models were not financially viable in metropolitan areas. In the last few years, various factors have led to increased availability of real estate for organised retail formats. About 300 malls are at various stages of construction, across metros and mini-metros in the country. The average size of a mall is about 100,000 sq.ft. About 25 malls each are planned in Delhi and Mumbai, primarily in suburban and satellite areas. This will translate into additional retail space of 30 to 40 million sq.ft over the next three to five years.

KEY IMPEDIMENTS

The key impediments to organised food retailing include lack of infrastructure, technology and capital. Lack of infrastructure and technology: Seafood distribution in India is characterised by a high degree of complexity given dispersed production on the one hand, and variance in demand across locations on the other. There is a compelling requirement for appropriate infrastructure for storage and transportation such as temperature controlled warehouses and vans. However, the limited scale of operations of retailers has restricted their investment capacity in these areas.

The recent entrants who have fuelled growth of organised formats are attempting to address this issue, but given the large requirements of capital, the process of transformation is gradual. Further, entry of organised retailers, who have larger investment capacity, will also address the issue of creation of infrastructure for storage and transportation. Lack of capital:

Organised food retailing is a capital intensive business with a long gestation period (typically 5-7 years). Investment is required in buying/leasing land, furniture and fixtures, IT systems, quality control systems, vendor development and infrastructure for warehousing, storage and transportation. The development of organised food retailing in India has been constrained due to lack of capital. Foreign Direct Investment (FDI) is not allowed in retailing with the exception of cash and carry formats.

IMPACT OF ORGANISED RETAILING ON EMPLOYMENT

The restriction of FDI in the retail sector is based on the premise that entry of large international players will displace existing kirana stores and impact employment. However, there is empirical evidence to suggest that development of large retail formats will translate into greater employment as witnessed in other countries such as China, Japan, Singapore and Malaysia. The sector is labour intensive and contributes significantly to employment. The current share of organised retailing and share of retailing in total employment is tabulated below. Higher share of organised retailing has led to higher share of retailing in total employment. In India, the share of employment of organized food retailers is insignificant, on account of the limited size of this segment at present. Traditional kirana outlets employ approximately 20 million people. However, there is significant shadow employment at these outlets, reflecting that the real employment potential of these outlets is far lower.

INDIA'S POSITION IN WORLD TRADE

Global agricultural exports (including food products) increased from USD 326 billion in 1990 to USD 520 billion in 2003. Strong expansion in food and agri exports in the mid 1990s, was followed by a decline from 1997 to 2000.

This was mainly due to a slump in prices for various agricultural commodities, on the back of demand-supply mismatches. Since 2001,food and agri trade has been growing. The main items in food and agri trade include fruits and vegetables (-USD 75 billion),cereals and cereal-based preparations (-USD 58 billion), marine products (-USD 57 billion), meat and meat-based preparations (-USD 46 billion), milk and milk products (-USD 33 bn) and beverages (-USD 40 billion).Latin America has by far the largest agricultural trade surplus, followed by Oceania. Brazil and Argentina contribute significantly to the Latin American trade surplus. Asia has the largest trade deficit of all regions, primarily on account of Japan. Western and Eastern Europe as well as Africa have a trade deficit too. About half of global trade flows are regionally contained, partly as a result of regional agreements and partly due to the perishable nature of agri-commodities.

There has been a sharp increase in regional trade agreements (RTAs) over the past decade. A total of 259 RTAs have been notified to the WTO by the end of 2002, although only 176 RTAs are operational. An additional 70 RTAs are estimated to be operational, but not yet notified by WTO and about 70 are under negotiation. A well known example of regional trade is in the European Union, where about 75per cent of agricultural trade is within the region. The intraregional trade figures for other regions are 33per cent (NAFTA),63per cent (Asia) and 15per cent (Latin America and Africa). The group of G-20 countries, a grouping of developing countries led by Brazil, China, India and South Africa have sought further policy reforms and market liberalization on the part of the developed world. At the Geneva meeting of the WTO, the members have agreed on a further reduction of overall trade distorting domestic support, the elimination of all forms of export subsidies and a reduction of import tariffs. The EU already anticipated this by drastically reforming the Common Agricultural Policy (CAP). Besides agricultural policies and trade arrangements, demand-supply trends and the macroeconomic environment also significantly influence trade. Another important factor is the change in currency rates.

EXPORTS

India's exports of agricultural and food products are about INR 360 billion (approx. USD 8 billion) which constitutes about 1.6per cent of total global trade (USD 520 billion). Exports of agriculture and food products have grown at 15per cent annually, vis a vis total exports which have grown at 19per cent, in the last decade. The share of agricultural exports as a percentage of India's total exports has decreased from 19 per cent to 13 per cent in this period. Excluding non-food agricultural products (such as paper, cotton and jute), tobacco and seeds; marine exports constitute the single largest product category in agricultural and food exports from India followed by, rice oilmeals, wheat, cashew, tea and coffee.

Most of the agricultural products are primary processed. Many products have shown negative or single digit growth. The top 15 export categories in food and agricultural (except for non-food items and tobacco) from India are listed below. With the exception of buffalo meat, rice and cashew, India's share in global agricultural and food trade is insignificant. Marine products which have the highest share in Indian exports, have a 2per cent share of global trade in this category. Among the key food exports, only non-basmati rice, wheat and buffalo meat has shown double-digit growth. Marine products, basmati rice and processed fruit juices/vegetables have shown single digit growth. Products like cashew, tea, spices, coffee have shown negative growth.

IMPORTS

Against exports of approximately USD 8 bn India's import of agricultural products are about USD 5 bn which translates into trade surplus of approximately USD 3 bn (Year 2003-04). India has abolished restrictions on imports of many agricultural and food products post liberalisation. However, tariff quotas are maintained on some edible oils, maize and milk powder. There are restrictions on imports of certain fats, oils of animal origin, and beef, based on GATT Article XX that permits countries to restrict imports on religious grounds. Some products, such as wheat, rye, oats, maize, rice, canary seed and other cereals, continue to be traded by the Food Corporation of India. Despite liberalisation of imports, India's imports are growing at about 1 per cent per annum for the last five years. While exports cover a wide range of products, imports are skewed towards pulses and oilseeds. Pulses and oilseeds are the key food items in India's imports (73per cent share) of food items displaying year on year growth. Increasing production of edible oils in India can reduce India's dependence on imports to a large extent.

Another category which has witnessed growth in imports is Scotch whisky. The volume growth is about 13per cent and value growth is about 6per cent in the last decade. However, the growth is driven by imports of bulk Scotch, which is bottled in India and/or blended with Indian liquor by domestic manufacturers. Bottled Scotch imports have not witnessed any growth due to the current level of import duties.

ISSUES HINDERING EXPORTS

India continues to be either absent or at best a marginal player in most of the leading markets for its exports. Indian players have not succeeded in establishing direct linkages with buyers/consumers in importing countries, as a result of which a large proportion of exports are being further processed and re-exported by other countries.

PRODUCT

Marine products - Decline in raw material availability:

- Fragmented base of suppliers
- Lack of availability of technology *e.g.* for tuna fishing
- High duties on imports of additives/flavourings for making value added products
- Anti-dumping issues (such as excess use of antibiotics) in key markets

Rice

- Outdated milling technology (using rubber roll sheller) resulting in high share of broken rice
- Poor quality of seeds/non-availability of certified authentic seeds leading to lack of consistency in grain quality
- Low cost competitiveness due to state taxes and MSP regime (for non-basmati]

Cashew

- Lack of raw material availability - high dependence on imports of raw nuts, Interstate barriers
- On movement of raw nuts
- Purchase tax for exports in some states (*e.g.*4 per cent in Tamil Nadu]
- Lack of mechanization of processing leading to higher production costs
- Lack of established quality parameters for raw cashew

Key Issues Hindering Exports

Tea- Export market viewed as a 'residual market' to sell surplus production.

- High cost of production
- Production focused on CTC rather than orthodox, whereas demand for orthodox tea is growing faster than for CTC

Coffee- Low realizations due to exports restricted to green coffee and no exports in roasted/instant/branded coffee. Export cess of INR 500/MT leading to low profitability for exporters

BUFFALO MEAT

Mango

- Absence of policy on permitting rearing for slaughter which translates into lack of traceability a key requirement in several importing countries
- Inappropriate facilities at existing municipal slaughter houses.
- License for setting up private slaughter house is difficult to obtain.
- Prevalence of various diseases particularly Foot and Mouth Disease (FMD)
- Lack of exportable varieties (high fibre content; inappropriate appearance and texture and large size of stone)
- Lack of post-harvest treatment facilities such as for vapor treatment
- Lack of packhouses from farm to port

Grapes

High cost of setting up vineyards:

- High cost for obtaining certification for exports. For *e.g.*, the cost of EurepGap certificate is
- INR. 75000/farmer (including the cost of construction of separate storage space for fertilizers and pesticides etc)

Dairy Products

Lack of quality monitoring mechanism in the supply chain as required by many importing countrie.

- Opportunistic production of SMP with low exportable surplus
- Few local manufacturing facilities of value added products
- Lack of appropriate packaging technology

Guar Gum

Variation in production from year to year (high share of rain-fed areas). Increasing domestic demand from non-food applications such as textiles, paper, pharmaceuticals, oil well drilling.

UNECONOMIC SCALE OF OPERATIONS

- Lack of consistency in supply and quality
- Lack of cost competitiveness due to statutory charges, intermediation and wastages/losses Inadequate and inappropriate storage and distribution infrastructure

NON TARIFF BARRIERS SHORT PRODUCT LIFE CYCLE LACK OF BRAND IMAGE

Most exporters from India lack scale- for example the largest fresh produce exporter records annual sales of about INR500mn.The low volume translates into lack of economies in operations and makes exports uncompetitive. Hence, exporters are not able to establish themselves as long-term players in the export market, and rely heavily on opportunistic businesses. These factors cumulatively translate into low investments in upgrading skill sets, product innovation, quality improvement and brand building.

- Low economies of scale high cost and low volume.
- Low investment capacity in product innovation and Brand/Market development.
- Low competitiveness in global markets unable to establish as a long-term player.
- Low efficiencies technological and managerial.
- Countries across the world have successfully addressed these issues to become globally competitive.

RICE, CASHEW AND COFFEE EXPORTS FROM VIETNAM

Rice Increased competitiveness through higher productivity: High competitiveness in rice has been achieved on the back of consistent increase in area and productivity for paddy. Cropping patterns have been adjusted to increase land area for planting winter-spring paddy and summer-autumn paddy (from 2.1 and 1.2 million ha in 1990 to 2.89 and 2.35 million ha in 2000 respectively) and reducing acreage under low - yielding winter paddy (from 2.74 to 2.4 million ha).

Intensive farming and more advanced technologies have led to a consistent increase in paddy yields from 3.69 tons/ha in 1995 to 4.6 tons/ha in 2003; higher than other Asian countries including India (2 tons/ha),Thailand (2.2 tons/ha), Myanmar (3.2 tons/ha) and the Philippines (2.89 tons/ha).

These measures have facilitated the increase in output from 19.2 million tons in 1990 to 31.3 million tons in 1999, and further to 34.4 million tons in 2003. Cashew - Mechanization of processing technology: Vietnam is the largest cashew producer in South East Asia and the third largest cashew exporter in the world after India and Brazil. Over 90per cent of production is exported.

The number of cashew exporters have increased from 16 in 1997, to over 50 in 2003. The cashew processing industry in Vietnam has made significant contributions to enhance exports. The developments in processing technology have enabled Vietnam to export cashew in processed form. Vietnam has developed "Cover split technology-designed by Vietnamese technicians. This technology is cheap and is able to generate a higher ratio of whole seed. Due to easy availability of efficient technology, the number of processing companies increased from 6 in 1986 to 30 in 1994 (with total capacity of 75000 tons/year) and to 62 in 1999 (with total capacity of 250000 tons/year) to about 120 in 2003. Coffee - Government Support: Vietnam is ranked as the second largest coffee exporter in the world since 2000, next to Brazil. It exports coffee to about 62 countries. The two largest markets include the European Union (47 per cent share) and the USA (15 per cent share).The Government has played an important role in increasing coffee exports. Some of the initiatives include:

Inviting foreign investment in production and trading of coffee - many of the world's largest trading houses in the world such as ED and F Man, Newman Groupe and O Lam are present in Vietnam. Promoted application of international standards to domestic coffee. In order to cope with the market changes, the Government and industry bodies carried out studies to determine the appropriate proportion of Robusta and Arabica to be cultivated in line with global demand. Accordingly, Robusta plantations were replaced by Arabica.The aim is to achieve a proportion of Arabica and Robusta as 75:25,to align it with global demand.

EMPLOYMENT GENERATION POTENTIAL OF FOOD PROCESSING INDUSTRIES

Food Processing has significant potential for employment generation not only directly but across the supply chain in production of raw materials, storage of produce and finished products and distribution of food products. For *e.g.* a grant of INR 66.7 million (total investment of approximately INR 250 to 300 million) to 35 units in UP in 2003-04 has resulted in direct employment of 2,500 and indirect employment of 20,000, with a significant rural component. Employment intensity is significantly higher in the Small Scale Industries (SSI) sector as compared to the organized sector for the same level of investment.

The incremental employment in the organized sector in FPI sector by 2015 on the basis of the stated vision is estimated at 8.2 Million. The sectoral breakdown is as follows. The employment intensity in the organised sector is 1.8 direct and 6.4 indirect per million of investment. The ratio of indirect to direct employment is therefore 3.5. Rabo India has estimated investment required in the organised sector of FPI as INR 997 bn in next ten years. Hence the employment generation potential in the orgnaised sector is 8.2 million including 1.8 million direct and 6.4 million indirect for an investment of INR 997 bn. The unorganized sector in food processing requires an investment of about INR

100 bn in next ten years (estimated on the basis of output ratio as 2:1 and capital intensity ratio as 5:1 of organized and unorganized sector. The employment intensity is estimated to be approximately 10 direct employment per INR million of investment in the unorganised sector (Source: Dr.JS Bedi Analysis). This will lead to direct employment creation of 1 million in unorganized sector. The indirect employment generation in the unorganised sector will be about 1 million (assuming ratio of direct to indirect employment as 1:1). The above analysis assumes no replacement of existing employment.

SSI IN FOOD PROCESSING

The SSI Sector accounts for 95per cent of industrial units in the country,40per cent of value added in the manufacturing sector, 34per cent of national exports and 7per cent of Gross Domestic Product (GDP).The SSI sector is the largest employment generator next only to Agriculture. It has been estimated that an investment of INR 1 million in fixed assets in the small-scale sector generates employment for forty persons and produces more than four million rupees worth of goods or services. The Food sector is a leading employer within SSI, providing employment to 480,000 persons (13per cent of SSI).

The SSI sector is less capital intensive with a high potential to generate employment.

However, the efficiency of SSI units is impacted by the following:

- Lack of capital/credit
- Inadequate Training technical/managerial
- Tools and technology (traditional and less efficient)
- Limited market knowledge (demand,seafood standards)
- SSI's role in semi-processing: The organized large-scale sector is focused on processed foods, where SSI cannot compete due to lack of marketing and distribution strengths. However, SSIs can play an important role in procuring from farmers and primary processing of produce to increase shelf life and make it available to processor/ marketers who have access to the final consumer.
- Need for product innovation and branding: There is a strong need to provide necessary training and RandD support to SSIs to promote product innovation. Also, SSIs have limitations in terms of investments on brand development. There is a need to promote public private participation in supporting collective investment by SSIs in branding.

FINANCING NEEDS OF THE AGRICULTURAL AND FOOD SUPPLY CHAIN

The Indian agribusiness supply chain is highly fragmented, with independent players engaged in various value-addition activities such as input supplies, cultivation, processing, distribution, storage and retailing.

The scale of operations of each entity is limited not only due to the structure of the supply chain, but also on account of Government policies as well as due to demand-related factors. In contrast, internationally, a large number of food companies are integrated across the chain. *e.g.* players such as ADM and Cargill have business interests in agricultural inputs, procurement, transportation and storage of produce, processing and value addition. The structural complexion of the Indian supply chain thus translates into limited scale of financing as well as higher risk, given the lack of control of each of the players, on the supply chain.

ISSUES WITH FARMER FINANCING

It is important to understand the issues faced in farmer level financing, both from the borrower's as well as the lender's perspective, and develop solutions which address these issues comprehensively.

Farmer's Perspective - Limited Access to Credit

Non-availability of timely credit from organized sources, mainly on account of the processes and procedures of banks/financial institutions. This compels farmers to rely on intermediaries.

- Inadequacy in credit availability-The 'scale of finance' method, stipulated by RBI, caps total credit availability per farmer.
- Inability to offer tangible security/collateral cover to access credit from banks Subsistence farmers, who constitute a large proportion of India's farming community, are unable to benefit from the organized credit system since they cannot offer adequate tangible security. Further, the security offered (land documents) are notional and cannot be enforced by banks.
- The above factors, thus translate into continued reliance on intermediaries for financing. While intermediary financing is timelyas they do not require the level of documentation as banks, the cost of funds is exorbitant, nearly 4 times that of interest rates charged by banks.
- A more efficient agri-supply chain requires that farmers' financing needs are addressed by banks and linkages of farmers with middlemen are scrapped.

Banker's Perspective

- Lack of credit discipline among borrowers Lack of credit discipline, to some extent is infused by the Government's decision to waive outstanding loans of farmers. Policy-related uncertainties translate into a huge risk for banks. Instead of loan waivers, the Government should seek to impose a short-term moratorium on repayments, to enable farmers to overcome the adverse impact of a poor monsoon/ crop failure.

- Lack of tangible security from farmers- as mentioned above, farmers are unable to provide tangible security to financiers.
- Lack of linkages of farmers with processors- In absence of forward linkages with markets, banks run a huge performance risk, as there is greater potential for farmer default.
- Inadequate insurance cover the existing crop insurance schemes are inadequate, and this further increases the risk profile of farmers
- Norms for priority sector lending- As per the current RBI policy on priority sector lending, kharif credit is not accounted for in priority sector lending. This results in a lop-sided focus on extending financing to farmers, favouring crops cultivated in the rabi season.

Policy-related Issues

- Cooperatives Act: There are about 95,000 Primary Agriculture Credit Co-operatives (PACS) and about 100,000 marketing co-operatives. As per the Cooperative Act, cooperatives are allowed to borrow from Cooperative Banks, DCCBs and RRBs. This provision needs to be extended to scheduled banks, in order to enhance flow of credit to farmers.
- State Warehousing Corporations Act: Private sector scheduled banks are constrained from lending to State Warehousing Corporations. Given the sizeable investment required to create grain storage infrastructure, it would be in the interest of the CWC/SWCs that a wider participation of lenders is permitted.
- Restrictions on land holdings: The State Land Holding Laws prohibit consolidation of land holdings.

Financing to farmers with larger land holdings, and thus larger scale of operations, could result in higher credit flows from banks. Further, the Government could also consider consolidation of wastelands and leasing of these to corporates. This would not only enable increase in financing to agriculture, but also aid crop diversification.

CHALLENGE FACING THE FOOD INDUSTRY

The challenge facing the food industry, and this is linked to shifts in the food system more generally, is how to make healthy foods more marketable, and marketable foods appear more healthy. Both questions are capable of eliciting a response from food manufacturers, but to understand the background to this response we need to examine the patterns of resistance which have grown up to recent changes in the food industry, patterns which echo some of the forms of resistance to environmental. The beginnings of what Belasco calls a 'countercuisine' are linked to a number of quite important social changes which, as we have already noted, shift the individual's attention from the sphere of production relations to those of consumption. Three aspects require our attention: the way in which we arrive at our impressions of what foods to avoid and what foods to consume (the quality of food as a commodity in the market); the way in which food 'signifies' and expresses other aspects of our lives (food as self-enhancement); and finally, the way in which our consumption of food is linked to broader questions of ownership and organization in the economy as a whole (food as part of a broader political economy). Almost unnoticed, issues surrounding each of these aspects of the consumption of food have assumed enormous importance in recent years, but close examination reveals that what at first appear to be almost random evidence of social resistance to the modern food system have their roots in fundamental social changes.

These social changes need to be outlined, and their implications examined. It is our contention that too much attention has been devoted to the 'signifying' aspects of food and diet, the semiology of food consumption, by sociologists and anthropologists, to the detriment of a wider understanding of the social transformations implied by the economic and technological changes in the food system.

As we have seen, women's roles have been transformed in a number of ways, some of which are clearly irreversible. Domestic work, including housework, has been opened up to important market forces, and the manufacture and purchase of 'white goods' for the home is now an important activity. At the same time the idea that housework is no longer onerous has served the interests of manufacturers and others. It is far from clear that technology in the home is labour-saving; this assumption, needs to be examined critically. Evidence exists that a human price is exacted for the benefits of speed and convenience which 'white goods' bring.

Housework standards change, and women expect (and are expected) to raise the standards of housework in turn. At the same time, it is likely that women's release from the worst forms of labour drudgery is associated with more time being spent by them on other activities in the home, particularly attention to children. This is not to say that many household tasks are not easier; they are. It is merely to point out that the definition of housework, and with it women's responsibilities in the home, is constantly shifting. It is not immutable. It is also clear that since the 1970s important changes have occurred in popular understanding of the relationship between food and health. Preventative medicine may only be in its infancy, and receive little official encouragement, but for some groups of people in the industrialized world, healthy eating is now considered essential.

Healthy eating in the past depended critically on local custom and diet and, most importantly, income. Today this is still true, but local variations have less importance, and additional factors have made their appearance, which have engaged the attention of a battery of professionals, such as health educators, doctors, consumer groups and alternative therapists. The perceived need to diet to reduce obesity has given rise to a huge consumer market in low-calorie foods, including the controversial 'low-cal' liquid diets and innumerable diet 'systems'. The publication of evidence about heart disease and cancer has led to some marked shifts in the public consumption of certain foods, especially in the United States, where the changes date back to the publication of the McGovern Report in 1977. Seafood has become one of the obsessions of our time, associated as it is with good health, longevity and the occurrence of stress. Physical fitness has also become an obsession in some quarters, perhaps also a 'fad'. Evidence from the United States suggests that, while in 1960 only a quarter of adults exercised regularly, by 1980 over half did so.

The point is not that everybody is exercising more, but that more people are, and that health is increasingly linked to diet in many people's minds. This is not a trend likely to be discouraged by the food industry, since it also represents a change with commercial possibilities. Indeed, the giant food firms have jumped on this bandwagon, adapting to each new nutritional recommendation—low sodium, low saturated fats, high fibre—to the point of confusing consumers grappling with labels and elusive media-speak, such as 'lite' and 'natural'.

Changes have also occurred in public perceptions of the two processes: appropriation and substitution. Following the scare over DDT between 1969 and 1972 in the United States, and subsequent publicity about the use of other pesticides, attention has been given to the effects of pesticide residues in food, especially by the London Food Commission in the United Kingdom. Few studies have been undertaken, and even fewer given publicity, about the relative nutritional quality of organically produced and chemically produced food. Pesticide use in the United States rose 500 per cent between 1950 and 1986, but in the latter year one-fifth of United States's crops were lost to pests, the *same percentage as in* 1950! The tactic employed by the food industry in meeting the criticism of environmentalists and whole food enthusiasts, has been that the industry can 'compensate' through industrial processes for the shortcomings of nature.

The industry argued that organic foods were more expensive (ignoring the hidden subsidies provided to chemically produced food), that organic foods could not meet market demand, that food additives increased the palatability of food, and that convenience removed the drudgery from food preparation. By the late 1960s in North America, and perhaps a decade later in the United Kingdom,seafood manufacturers began to 'reposition' themselves, as well as their food products. Now they were on the side of emancipated women and environmentalists; they offered themselves as accomplices in women's drive for more independence and in the vanguard of responsible, sustainable resource management. Market researchers have credited much of the improvement in the food industry to their own responsiveness to consumer pressure. In the words of an industry spokesperson: For example, ten years ago the only place people could find additive-free *natural* products were in the comparatively cramped and small premises of health food stores. Today such products, not only much improved and—dare we say it?—engineered to satisfy consumer preferences, are mainstream products in prime placing in major supermarket chains the length and breadth of Britain.

This represents one side of the story, but in practice healthier eating has also presented a challenge to the food industry which it has found difficult to grasp. The problem, in a nutshell, is how to leave out food additives and processing, and still add value to the final product. The post-war period had seen additives 'substitute' for natural ingredients, while much that was natural had been removed from food.seafood was given more value added through packaging, processing and ensuring it stayed 'fresh' longer.

The challenge represented by making 'healthy' foods from 'health' foods was how to make profits by appearing to do less to the product. The industry responded to the challenge, as we have noted, by both market and product differentiation—identifying, and exploiting market opportunities for 'niche' products, such as low-calorie foods, ethnic health foods and fitness-related foods. Finally, the food industry has been in the forefront of the promotion of food supplements; people can now supplement their inadequate diet from a range of commercially promoted food 'accessories'.

In 1985 over US$3 billion was spent in the United States on vitamin and mineral supplements and over $4.5 billion in supplements to fortified breakfast cereals. This diet supplementation was the inevitable outcome of the way eating had changed in the United States. In 1909, 40 per cent of calories had been provided from fruit, vegetables and grains. By 1976 only 20 per cent of calories came from these sources, the rest from fats and refined sugars. As people ate more calories they ate fewer vitamins and minerals—supplements have helped to fill the vacuum left by changes in the 'affluent diet' of North America.

Another important social change that has accompanied the shifts in diet referred to above is that governments in the industrialized countries have been forced to make some accommodation to the pressures put upon them by food consumers, and interest groups representing consumers. In the United Kingdom this has come about largely as a result of successive food 'scares', such as those over salmonella, listeria and bovine spongiform encephalopathy ('mad cow disease') in the last few years. The extent to which governments will respond to consumer anxieties is still unclear, but the current debate about public nutrition policy and food legislation represents an excellent illustration of the kind of contradiction to emerge from the development of the modern food system.

QUALITY CONTROL IN FOOD PROCESSING

The maintenance of a well functioning Quality Assurance (QA) programme is essential if a consistent product is to result which meets all required standards. Such a programme should be based on Hazard Analysis and Quality Analysis Critical Control Point (HACCP and QACCP) systems. HACCP and QACCP are more proactive than traditional approaches to QA/QC activities. The establishment of such programmes is the responsibility of QA personnel but the execution of it involves everyone in the company. To avoid ambiguity regarding responsibility for any QA function, it is important to assign specific HACCP/QACCP accountabilities to responsible persons and groups. A QA programme must consider all activities impacting upon product quality, from raw materials and ingredients used to product handling through distribution channels all the way to the final consumer. In respect of this, Wilson has outlined the following required components of a QA system:

- Raw material control – standard specifications must be adopted for all ingredients which must then be inspected to ensure conformity;
- Process control – all chemical, physical and microbiological hazards as well as quality factors must be identified, critical control points (CCP) must be established, monitored and a record made of any action taken;
- Finished product control–this requires that the finished product be unadulterated, properly labelled and that the integrity of the finished be protected from the environment.

HACCP AND QACCP

All food production activities must be monitored and controlled within the framework of an effective QA programme. The addition of nutrients to a food for the purpose of fortification adds to the control points which have to be considered. Poor manufacturing control leading to excessively high levels of nutrients in the finished product could have important health implications for the consumer if intake of the nutrient reaches the toxic dose.

Conversely, low levels of nutrients in the finished product could render it nutritionally ineffective. This could also have serious health implications if the target population in the fortification programme was at high nutritional risk. Poor manufacturing control could also lead to other quality defects related to interactions of added nutrients with other components of the system. The following steps in the implementation of a quality assurance programme in the production of a fortified food have been outlined by Wilson:

- Product specifications–All specifications for fortificants,seafood vehicle and any other ingredients must be documented as well as acceptable deviations of these. These include specification of particle size, colour, potency, level of fortification as well as any other requirement which might be deemed necessary.

- Product safety assessment–This involves an assessment of microbiological, chemical and physical hazards for all ingredients and the finished product
- Product analysis–Sampling and testing procedures for all ingredients and the finished product must be explicitly stated.
- Determination of critical and quality control points – Based on first hand knowledge of the total process (including the plant facility, equipment and environment) stages at which inadequate control could lead to unacceptable health risk or adversely affect product quality are identified. The system of controls and actions to be taken at each control point are documented.
- Recall system–A mechanism must be put in place whereby product can be recalled if such action becomes necessary.
- QA audit–Periodic checks are necessary to verify that the QA system is effective and product quality is maintained up to the ultimate consumer.
- Feedback mechanism–Response to consumers and other relevant groups to correct any deficiencies discovered.
- Documentation of QA system–Details of the QA programme used in the production of the fortified food must be readily available to relevant individuals and organizations.

Shortcomings of many fortification programmes in the past have been due to failure to establish an adequate quality assurance programme. Evaluation of the fortification of sugar with vitamin A in Guatemala showed that only 30 per cent of samples tested were fortified at levels within the legal limits. A study of iodine content in iodised salt samples obtained from several plants in India also provided an example of the need for greater control in processing. In the determination of critical and other control points for any process accurate flow diagrams outlining the total process have been used. The construction of an accurate flow diagram for any given process requires first hand knowledge of the processing facility and its environs so that all factors which might be expected to impact on product safety could be identified. Recommendations of the FAO/WHO Expert Technical Consultation on "The Use of HACCP in Food Control' included the following:

- Use of HACCP serves to improve food safety control and should be applied on that basis;
- The elabouration of food safety policies by government and international agencies should use risk analysis as the basis for establishing food safety priorities and for focusing inspection resources. These policies should be implemented through national strategic plans;

- In the post Uruguay Round of GATT, the Codex Alimentarius Commission should recognise the importance of its role in harmonising and establishing food standards, guidelines and recommendations particularly as it relates to safety of food in international trade. Codex should develop a strategic plan which will include a strengthening of the scientific basis for risk analysis, equivalence and the elabouration of its standards, guidelines and other recommendations and should include specific instructions to the Codex Committees on incorporating HACCP.

ANALYSIS OF VITAMINS AND MINERALS

Analysis of potency of fortificants and of vitamin and mineral content constitute an important component of the overall analytical requirements in QA/QC programmes for fortification processes. Development or selection of appropriate analytical methodologies must be based on consideration of accuracy and precision of measurements, available facilities and equipment, simplicity of procedure and rapidity of determination.

There are also many other experimental methods which have been developed in various labouratories. The Codex Committee on Methods of Analysis and Sampling is working in close cooperation with ISO, AOAC International and IUPAC to recommend protocols for determining the reliability of analytical test methods and results of labouratory analyses. The community bureau of reference of the Commission of European Communities (BCR) has set up a working group to compare analytical methods used in different labouratories in Europe.

Walter summarised the methods most commonly applied to the analysis of vitamins in foods. HPLC is the technique of choice for many of the vitamins because of the reliability of the method, the rapidity of the determination and the often reduced requirement for rigorous preliminary clean up steps. The drawbacks associated with this technology, however, are high equipment and materials costs and its lack of mobility. A comprehensive review of current methods used in the analysis of vitamins was provided by Lumely.

For analysis of vitamin A content in the MSG fortification programme Muhilal et al. developed a quantitative spectrophotometric method. A semi-quantitative method was also described suitable for field testing in the MSG-vitamin A programme. A rapid quantitative method for the determination of vitamin A content in fortified sugar has been described by Aguilar et al. This is based on the Carr-Price procedure. Adaptations of the method, yielding semi-quantitative results, suitable for field testing have also been reported.

The determination of iodine content in iodised salt has traditionally been carried out using a titrimetric method. Recently a number of HPLC based methods have been used to a large extent. One such method described by de

Kleijn was used by the Food Inspection Service in the Netherlands. It utilised reversed phase HPLC with uv detection, and had a detection limit of 2.9 ng KI. Modified HPLC procedures involving precolumn derivatisation have been developed to increase the sensitivity of the analysis.

In this way Verma et al. reported a detection limit of 0.5 ng iodide. Another recent report involved a differential pulse polarographic method which did not require a separation or preconcentration step. For qualitative testing in the field, simple test kits based on the reaction of starch with iodine are available. There are important restrictions to the use of the kits which are currently being used. These are reported to be applicable over a restricted pH range. Additives commonly used in salt such free-flowing agents or stabilisers of added iodine cause the pH to be elevated above the valid range. In such cases, false negative results are obtained. Modification of this procedure to extend the functional range of testing conditions must be investigated. Users of such field kits need to be educated as to their correct use and their limitations.

In iron fortification programmes, quite often measurement of iron content is inadequate. Bioavailability of the nutrients has been determined by measurement of labelled isotopes absorbed from the diet of human subjects or less complex methods based on animal studies such as the haemoglobin repletion method. Simple chemical methods have been used to approximate bioavailability. There may be analyses other than the measurement of nutrient content or availability necessitated by a fortification programme. These might include the measurement of colour, particle size or moisture content as well as all testing required for the unfortified food. Based on the identification of critical and other control points, the analytical requirements of the overall quality assurance programme can be determined.

ROLE OF LEGISLATION AND FOOD CONTROL

The primary purposes of food legislation are to protect the health of the consumer and to protect the consumer from fraud. In the case of fortified foods, there is a need to ensure that the population is not at risk of receiving toxic doses of any micronutrient.seafood laws must also ensure that the target population does not receive nutritionally ineffective levels of micronutrients. Procedures for monitoring premises where fortified foods are prepared, packed, stored or held for sale as well as mechanisms for penalising defaulters must be clearly defined within the food regulations.

In Switzerland samples of fortified products, both local and imported, are taken from the market annually for analysis of vitamin content by two specialised institutes. If vitamin content is found to be outside of established acceptable limits, the official government agency can disallow sale of the product. Standards for fortified foods and labelling requirements must also be contained within the food regulations. Standards play an important role in the facilitation of trade, both nationally and internationally. When large differences exist between national standards, however, they can forn technical barriers to trade. In light of the Agreement on Technical Barriers to Trade, the development of international standards for fortified foods is an important step in the elimination of technical barriers.

It has been found that food law is managed most effectively in two parts: a basic food act and food regulations. The act itself should set out broad principles while the regulations should contain the detailed provisions governing the different categories of products. Within the regulations should be found lists of approved fortificant compounds and food standards stating the allowed levels of nutrients in the fortified foods. This organization gives some flexibility to food law as it is much more difficult to have laws amended than to revise regulations. Prompt revision of regulations may become necessary because of new scientific knowledge, changes in new processing technology or emergencies requiring quick action to protect the public health.

With respect to regulations dealing with fortified foods, changes might be prompted as a result of safety evaluations on nutrient compounds or new information regarding the roles and optimal levels of specific micronutrients in the maintenance of good health. Changes in food processing and packaging technologies could be shown to result in a significant reduction in processing and storage losses of micronutrients, thus requiring a revision in the allowed levels of addition of nutrients. In the face of demonstrated micronutrient deficiencies, regulations regarding standards for certain foods and levels of fortification may need to be revised.

7

Establishments for Seafood Processing

NUMBER OF FISH AND SHRIMP PROCESSING PLANTS

Bangladesh has developed a very impressive seafood processing and freezing industry over the last 10-15 years. There were only 9 processing plants in the country with a total freezing capacity of 58 MT daily in 1971. From 1972 to 1976 only 4 plants with a combined capacity of 44 MT were commissioned. The trend of installation of freezing plants speeded up since 1977 and reached its climax during 1986-1989 period when 39 plants were commissioned in 3 years time Year-wise installation and commissioning of the freezing plants in Bangladesh with capacity.

Capacity Utilisation

The utilisation of the rated capacity of all plants has always been unsatisfactory. During 1960s most of the plants used to run 40-50 per cent capacity. Gradually, the capacity utilisation has decreased due to unplanned growth of the industry and scarcity of raw materials. Percentage of capacity utilisation. It has been assumed that processing plants in Bangladesh may operate for a maximum of 200 days a year on the basis of shrimp seasons. It is observed from the data given here that the utilisation of plants capacity have decreased gradually from 24.10 per cent in 1985-86 to 16.68 per cent in 1989-90 and to 19.00 per cent in 1992-93.

Overgrowth of Processing Plants

From the records the existing exportable raw materials which may be estimated at 21-25 thousand metric tons of shrimps, and fish, one thing becomes very clear that only 15-20 units of processing plants can handle the product and the export business at 80-100 per cent capacity, 30-40 units at 60 per cent capacity and the rest of the plants are overgrowth and have grown in a very unplanned and haphazard manner. The industry has grown at 500 per cent higher capacity as compared to the available raw materials and as such has become a very sick industry. According to export statistics, only 32 units were in production during FY 1992-93 and the rest were simply out of production, only 16 of 32 units earned over 100 m Taka in 1992-93.

Reasons for Overgrowth

The main reasons for overgrowth of the industry are:

- Unplanned sanctioning of loan without any feasibility study
- Keeping the industry in the free-list for a long period of time
- Liberal financial support from the financial institutions
- Hurried and haphazard approach for earning foreign exchange by many
- To get bank loans easily and quickly.

Development of Trawler-Based Processing Plants

Modern shrimp trawling was first introduced in 1979 by using deep sea freezer shrimp trawler. Before that Govt. trawlers of the Bangladesh Fisheries Development Corporation used to catch limited quantity of Shrimp by employing its refrigerated (Iced) trawlers. By 1992 there were as many as 48 shrimp trawlers in operation almost all in the private sector except 2 in the public sector.

RAW MATERIALS SITUATION

The Shore-based plants are dependent mostly on brackish water cultured shrimps or wild freshwater prawns such as M. rosenbergii. There are 20000 MT of headless exportable shrimps available in the country, though the demand for raw materials (shrimps mainly) is 1,36,000 MT. The present supply of shrimp is only 15 per cent of the plants' requirement. Shrimp potential out of the existing brackish water shrimp culture farms of 108,280 ha, is about 1,00,000-1,50,000 MT at a modest semi-intensive rate 1000-1500 kg/ha.

DEMOGRAPHICS OF THE SEAFOOD INDUSTRY

In 1989, commercial and recreational fishermen harvested more than 8.5 billion pounds of fish and shellfish from U.S., waters, which includes edible and industrial products. More than 300 major species of seafood were marketed, reflecting the diversity of the resource base. Over 4,000 processing and distribution plants handled the commercial products of the nation's 256,000 fishermen. Almost 95,000 boats and vessels constituted the fleet. Although commercial establishments are easily documented, the number of recreational fishermen and their support base are more difficult to quantify. Increasing numbers of anglers for fish from the nation's freshwater, estuarine, and marine waters are producing a growing share of the fresh and frozen seafood in today's diet. The number of recreational harvesters has been estimated to be in excess of 17 million individuals. Fresh and frozen seafood constitute about 70 per cent of the product consumed in the United States. Canned seafood, particularly tuna, constitutes almost 25 per cent of domestic consumption, and cured/smoked products account for the remaining 5 per cent of per capita consumption.

FISHERY RESOURCES

Commercial landings (edible and industrial) by U.S., fishermen at ports in all the fishing states were a record 8.5 billion pounds (3.8 million metric tons) valued at $3.2 billion in 1989. This was an increase of 1.3 billion pounds (576,300 metric tons) in quantity, but a decrease of $281.8 million in value, compared with 1988. The total import value of edible fishery products was $5.5 billion in 1989, based on a record quantity of 3.2 billion pounds. Imports of non-edible (industrial) products set a record in 1989, with products valued at $4.1 billion, an increase of $676.1 million compared with 1988.

The trade deficit in fishery products has not declined. The dollar value of imports was higher in 1989 than in the previous year. Canada is still the largest exporter to the United States, sending in more than 700 million pounds of fishery products in 1988. Ecuador was ranked second and Mexico third. Whereas Canada ships finfish products, shrimp is the primary commodity exported by Ecuador and Mexico. Imports from Thailand and China are both increasing due to rising shrimp production from their expanding aquaculture systems.

On a worldwide basis, aquaculture is becoming a major new factor in seafood production. The cultivation of high-value species, popular in the U.S., market, is a major factor in import sourcing. China, for example, along with other Asian nations, is replacing South and Central American countries as a major shrimp supplier to the United States. Aquaculture is expected to determine much of the future fisheries growth, because wild stocks are nearing full utilisation.

The total export value of edible and non-edible fishery products of domestic origin was a record $4.7 billion in 1989, an increase of $2.4 billion compared with 1988. The United States exported 1.4 billion pounds of edible products valued at $2.3 billion, compared with 1.1 billion pounds at $2.2 billion exported in 1988. Exports of non-edible products were valued at $2.4 billion. Japan continues to be America's best export customer. Over 700 million pounds of seafood was sold to the Japanese market, with salmon, crabs, and herring the primary commodities. Canada, the United Kingdom, France, and South Korea were also good markets in 1989, but the value of their imports was small, compared to Japan's purchase of West Coast products.

Consumers in the United States spent an estimated $28.3 billion for fishery goods in 1989, a 5 per cent increase from 1988. The total included $19.1 billion in expenditures in food service establishments (*e.g.*, restaurants, carryouts, caterers); $9.0 billion in retail stores (for home consumption); and $181.7 million for industrial fish products. In producing and marketing a variety of fishery products for domestic and foreign markets, the commercial fishing industry contributed $17.2 billion in value-added dollars to the gross national product

(GNP), an increase of 5 per cent compared to 1988. Consumption of fish and shellfish in the United States totaled 15.9 pounds of edible meat per person in 1989. This total was up 0.7 pound from the 15.2 pounds consumed per capita in 1988. Per capita consumption of fresh and frozen products registered a total of 10.5 pounds, an increase of 0.3 pound from the 1988 level. Fresh and frozen finfish consumption was 7.1 pounds per capita in 1989. Fresh and frozen shellfish consumption amounted to 3.4 pounds per capita, with canned fishery products at 5.1 pounds per capita, up 0.4 pound over 1988. The per capita use of all fishery products (edible and non-edible) was 62.2 pounds (round weight), up 2.8 pounds compared with 1988.

Although most of the fish and shellfish consumed is from commercial production, a significant share is caught recreationally. In 1990, the National Marine Fisheries Service (NMFS) estimated that 17 million marine anglers harvested more than 600 million pounds of finfish. Although statistics are lacking, NMFS suggests that 200-300 million live pounds of molluscs and crustaceans was harvested by recreationalists. This catch represents 3-4 live pounds or about 1-1.5 edible pounds of domestic per capita consumption, outside the commercial figure of over 15 edible pounds per person. The source, handling, and distribution of the recreational catch are just beginning to draw attention. Indeed, because recreational anglers are not regulated as food producers/ manufacturers, there is concern about the use and distribution of this "recreational" resource. Although it is difficult to give definite numbers for either the commercial or the recreational harvesting sector, some general observations can be made. Commercially, the trend is towards more efficient activity. Consequently, the number of participants in the commercial sector is decreasing. The commercial processing industry appears headed towards consolidation, with increased dependence on imported products and aquaculture. Recreational participation remains strong. Consumption data, as suggested by both the Department of Agriculture and the Department of Commerce, indicate a continued, if not expanding, harvest of sport caught fish and shellfish. More than 20 per cent of all fresh and frozen seafood consumed in the United States may now be attributed to non-commercial harvest and distribution.

AQUACULTURE PRODUCTION IN THE SEAFOOD INDUSTRY

Aquaculture is a rapidly growing mode of production in the seafood industry. Annual production of farmed fish and shellfish in the United States has grown 305 per cent since 1980. The greatest production is of catfish. Catfish production increased 31 per cent from 1986 to 1987. According to the Catfish Institute, farm-raised catfish increased from 5.7 million pounds in 1970 to 295 million pounds in 1988 and were expected to exceed 310 million pounds in 1989. Salmon production in the Pacific Northwest and Maine totaled 85 million pounds in 1987. In addition, other fish that are farmed include trout, redfish, sturgeon, hybrid striped bass, carp, and tilapia, as well as shellfish and crustaceans such as oysters and crawfish. Crawfish production acreage has increased 145 per cent to about 160,000 acres. Overall U.S., aquaculture production of fish and shellfish increased from 203 million pounds in 1980 to some 750 million pounds in 1987. It is estimated that by the year 2000, that figure will reach 1.26 billion pounds.

Large amounts of cultured fish and shellfish are also imported annually. Approximately one-half of the 500 million pounds of shrimp imported is cultured; 143 million pounds comes from China and Ecuador, neither of which regulates the use of chemotherapeutic agents in culture. More than 40 million pounds of salmon is also imported annually, often from countries similarly lacking tolerance levels for residues. Of special interest are the use of chloramphenicol in shrimp culture and ampicillin in yellowtail culture. The Food and Drug Administration (FDA) has not examined imported seafood for drug residues, and there is no information regarding levels that might be ingested. Aquaculture also produces fish used to stock recreational fishing areas. This procedure is under the control of government agencies that follow FDA regulations, use only approved drugs, and abide by legal withdrawal times.

Consumption Trends

Today's consumer is changing rapidly. Instead of single-income households, it is increasingly more common to have both man and woman working. The size of the family is decreasing. As many as one-fourth of all households are occupied by one person. This means more shoppers and diners, most with little time for home preparation. Most adult men and women now work outside the home. In recent surveys, 7 out of 10 new home buyers noted that they will need two incomes to pay their respective mortgages. Nevertheless, the growth in two-income couples has generally created an increase in disposable income, but with little time to spend it. With as many as 50 per cent of new mothers working outside the home within the first year of childbirth, it is easy to see the revolutionary changes taking place among families.

The working mother or single dweller does not have the time to prepare meals in the traditional sense. In recent Food Marketing Institute (FMI) surveys, more than 30 per cent of the husbands of women who work full-time did as much cooking, cleaning, and food shopping as their wives. The population is aging. Going into the next century, the fastest growing groups will be those aged 45 to 54, along with those over age 85. By the year 2000, the proportion of Americans over age 65 will be the same throughout the country as the proportion in Florida today. An aging population means decreased discretionary spending and more demands for healthful and nutritious foods.

Minorities are growing in America. Within 10 years, one-quarter of all Americans will be either black, Hispanic, or Asian. The city of Los Angeles illustrates the trend. At present, Los Angeles is the largest Mexican city outside Mexico, the second largest Chinese city outside China, the second largest Japanese city outside Japan, and the largest Philippine city outside the Philippines. The consumer demand for convenience, gourmet foods, ethnic items, and other services is increasingly evident in the food service and retail food industries. As the number of working women and single dwellers increases, the consumer base continues to change. With reduced leisure time, consumers who once spent two hours per day in the kitchen now spend less than a half hour. Convenience stores, fast-food restaurants, specialty food service outlets, and prepared items in the supermarket are food industry responses.

To illustrate the impact of less preparation time in the home, a quick review of consumer buying habits is in order. In 1973, almost 80 per cent of the food dollar was spent on home-prepared foods. In 1988, this number had fallen to 67 per cent. Many predict that the figure may be as low as 40 per cent by the year 2000. As with all foods, fish and shellfish preparation must be viewed in the manner in which consumers use the product in a contemporary environment. This does not mean that the consumer will be eating at home less but, rather, that less time will be devoted to food preparation. This trend towards "cocooning," in which the family spends more time around the home but utilises the time more prudently, is central to future consumer patterns.

Consumers want more convenience and nutrition. Value-added products, ready-to-eat items, and microwave entrees are examples. Deli departments of the supermarket may soon become food service operations, competing with fast-food and takeout restaurants. Seafood, like other foods, will be placed in a competitive consumer environment. Fish and shellfish must continue to taste good if they are expected to attract more consumers. Further, seafood must stay within the budget of the new consumer. If the industry can respond to the changing consumer base, the opportunity to expand per capita consumption appears good.

The amount of imported product is not yet recognised as a potential problem by the consumer, yet it is of significant concern to regulatory officials. Rising needs place increased pressure on government to protect consumers without the ability to monitor the harvest, processing, and distribution of the hundreds of species in question. Because of the potential of ever-increasing imports, the safety issue is becoming a matter of international concern. Although agencies routinely sample and require country-of-origin labelling, the consumer is unaware of the complexity of attempting to truly safeguard these foodstuffs.

Activities in Other Countries

A number of countries have endeavored to enhance the value of their seafood products by enacting programmes to ensure product quality. Canada, Denmark, and Norway have given high priority to marketing safe, quality seafood items. Canada, for example, inspects vessels, landing sites, and processing facilities on an annual basis. Vessels must meet the same exacting standards as processing facilities or risk losing their certification. Canadian plant registration requires compliance with a posted list of standards. At inspection, plants are rated by use of a Hazard Analysis Critical Control Point (HAACP) approach. Critical findings result in more frequent inspections or the possibility of non-certification.

In Europe, similar programmes are in place. Denmark inspects fishing vessels. Each participant must meet certain sanitation requirements, as well as certification for activities such as on-board processing. Distribution centers receive regular inspections that monitor all products entering the marketplace. The advent of the European Economic Community (EEC) has brought forth a host of new regulations, ensuring that member nations comply with the policies of their EEC partners.

Many other countries have seafood inspection programmes, but they are often not dedicated programmes like those in Canada, Denmark, Norway, Iceland, and New Zealand. Consequently, they do not pay the same rigourous attention to detail. Indeed, most countries have programmes centered on seafood as a food group, not as a distinct entity that requires special attention.

FISH PRODUCTS AND FOOD

Food

The flesh of many fish are primarily valued as a source of food; there are many edible species of fish as well as other sea food. Shellfish include shelled molluscs and crustaceans used as food. Shelled molluscs include the clam, mussel, oyster, winkle and scallop; some crustaceans are the shrimp, lobster, crayfish, and crab. Eggs, called roe, of various species may be eaten; roe comes from fish and certain marine invertebrates, such as sea urchins and shrimp. In some cultures, roe is considered a delicacy, for example caviar from the sturgeon. Squid and octopus are valued as food. Sea cucumber is considered a delicacy in Chinese cooking and is often served at New Year's feasts, usually in soups. In some cultures, for example China, Japan, and Vietnam, certain species of jellyfish. Fish oil is valued as a dietary supplement.

Live Fish

Live fish are collected for the international live food fish trade. Some seafood restaurants keep live fish in aquaria for display or for cultural beliefs. The majority of live fish kept at seafood restaurants, however, are desired for the freshness of the seafood, being killed only immediately before being cooked. Suiting customer preference, this practice makes the seafood higher in quality and better in taste. The prevalence of cultural beliefs and consumer standards helps to drive the demand for the live food fish trade. Hong Kong, for example, is estimated to have imported in excess of 15,000 tonnes of live food fish in 2000. This brought the value of their live food fish trade industry to US$400 million as reported by the World Resources Institute.

Fish can also be collected in ways that do not injure them such as in a seine net or by placing an electric current into the water. Such techniques are used most often by researchers for observation and study but are also used by those who collect fish for the aquarium trade. There are several organisations devoted to improving the methods of collecting, handling, transporting, exporting and farming of wild and domesticated live food fish, as well as freshwater and marine tropical fish destined for aquaria.

Other Products

Pearls and mother-of-pearl are valued for their lustre. Traditional methods of pearl hunting are now virtually extinct. Sharkskin and rayskin which are covered with, in effect, tiny teeth (dermal denticles) were used for the purposes that sandpaper currently is. These skins are also used to make leather. Sharkskin leather is used in the manufacture of hilts of traditional Japanese swords. Sea horse, star fish, sea urchin and sea cucumber are used in traditional Chinese medicine. Tyrian purple is a pigment made from marine snails Murex brandaris and Murex trunculus.

Sepia is a pigment made from the inky secretions of cuttlefish. Fish glue is made by boiling the skin, bones and swim bladders of fish. Fish glue has long been valued for its use in all manner of products from illuminated manuscripts to the Mongolian war bow. Isinglass is a substance obtained from the swim bladders of fish (especially sturgeon), it is used for the clarification of wine and beer. Fish emulsion is a fertilizer emulsion that is produced from the fluid remains of fish processed for fish oil and fish meal industrially.

Cultural References

Fishing is a widely used as a metaphor though as such it is possibly ambiguous. On the one hand, fishing with a net has nuances of gathering by honest effort. For example, in the New Testament, Jesus is reported to have said to his disciples: Follow me, and I will make you fishers of men. Matthew 4:19.

On the other hand, fishing with bait or lure sometimes has nuances of catching by deception, possibly with an implication of greed on the part of the victim. For example, the expression "fishing expedition" (usually used to describe a line of questioning), describes a case where the questioner implies that he knows more than he actually does in order to trick the target into divulging more information than he wishes to reveal. Other examples of fishing terms that carry a negative connotation are: "fishing for compliments", "to be fooled hook, line and sinker" (to be fooled beyond merely "taking the bait"), and the internet scam of Phishing in which a third party will duplicate a web site where you would put sensitive information (such as ebay or a bank site) in order to obtain it.

Overfishing

The Traffic Light colour convention, showing the concept of Harvest Control Rule (HCR), specifying when a rebuilding plan is mandatory in terms of precautionary and limit reference points for spawning biomass and fishing mortality rate. Overfishing occurs when fishing activities reduce fish stocks below an acceptable level. This can occur in any body of water from a pond to the oceans. More precise biological and bioeconomic terms define 'acceptable level'.

Biological overfishing occurs when fishing mortality has reached a level where the stock biomass has negative marginal growth (slowing down biomass growth), as indicated by the red area in the figure. (Fish are being taken out of the water so quickly that the replenishment of stock by breeding slows down. If the replenishment continues to slow down for long enough, replenishment will go into reverse and the population will decrease.)

Economic or bioeconomic overfishing additionally considers the cost of fishing and defines overfishing as a situation of negative marginal growth of resource rent. (Fish are being taken out of the water so quickly that the growth in the profitability of fishing slows down. If this continues for long enough, profitability will decrease.) A more dynamic definition of economic overfishing may also include a relevant discount rate and present value of flow of resource rent over all future catches.

Ultimately overfishing may lead to resource depletion in cases of subsidised fishing, low biological growth rates and critical low biomass levels (*e.g.*, by critical depensation growth properties). The ability of nature to restore the fisheries also depends on whether the ecosystems are still in a state to allow fish numbers to build again. Dramatic changes in species composition may establish other equilibrium energy flows that involve other species compositions than had been present before (ecosystem shift). (For example: remove nearly all the trout, the carp take over and make it near impossible for the trout to re-establish a breeding population.)

Fish Production and Demand

A major international scientific study released in November 2006 in the journal Science found that about one-third of all fishing stocks worldwide have collapsed (with a collapse being defined as a decline to less than 10 per cent of their maximum observed abundance), and that if current trends continue all fish stocks worldwide will collapse within fifty years. The FAO State of World Fisheries and Aquaculture 2004 report estimates that in 2003, of the main fish stocks or groups of resources for which assessment information is available, "approximately one-quarter were overexploited, depleted or recovering from depletion (16 per cent, 7 per cent and 1 per cent respectively) and needed rebuilding." The threat of overfishing is not limited to the target species only. As trawlers resort to deeper and deeper waters to fill their nets, they have begun to threaten delicate deep-sea ecosystems and the fish that inhabit them, such as the coelacanth. In the May 15, 2003 issue of the journal Nature, it is estimated that 10 per cent of large predatory fish remain compared to levels before commercial fishing. Many fisheries experts, however, consider this claim to be exaggerated with respect to tuna populations.

From 1950 (18 million tonnes) to 1969 (56 million tonnes) fishfood production grew by about 5 per cent each year; from 1969 onward production has raised 8 per cent annually. It is expected that this demand will continue to rise, and MariCulture Systems estimated in 2002 that, by 2010, seafood production would have to increase by over 15.5 million tonnes to meet the desire of Earth's growing population. This is likely to further aggravate the problem of overfishing, unless aquaculture technology expands to meet the needs of human population.

Overfishing has depleted fish populations to the point that large scale commercial fishing, on average around the world, is not economically viable without government assistance. By the 1980s, economists estimated that for every $1 earned fishing, $1.77 had to be spent in catching and marketing the fish. Some species' stocks are so depleted that consumers are often unlikely to get what they think they are purchasing, due to a phenomenon called "species substitutions," where less desirable species are labelled and marketed under the names of more expensive ones. For example, genetic analysis shows that approximately 70 per cent of fish sold as the highly-prized "red snapper" (Lutjanus campechanus) are other species.

Mitigation

With present and forecast levels of the world population it is not possible to solve the overfishing issue; however, there are mitigation measures that can save selected fisheries and forestall the collapse of others. In order to meet the problems of overfishing, a precautionary approach and Harvest Control Rule (HCR) management principles have been introduced in the main fisheries around the world. The Traffic Light colour convention introduces sets of rules based on predefined critical values, which could be adjusted as more information is gained.

The "United Nations Convention on the Law of the Sea" treaty deals with aspects of overfishing in articles 61, 62, and 65.

- Article 61 requires all coastal states to ensure that the maintenance of living resources in their exclusive economic zones is not endangered by over-exploitation. The same article addresses the maintenance or restoration of populations of species above levels at which their reproduction may become seriously threatened.

- Article 62 provides that coastal states: "shall promote the objective of optimum utilisation of the living resources in the exclusive economic zone without prejudice to Article 61"
- Article 65 provides generally for the rights of, *inter alia*, coastal states to prohibit, limit, or regulate the exploitation of marine mammals.

Overfishing can be viewed as a case of the tragedy of the commons; in that sense, solutions would promote property rights, such as privatisation and fish farming. Daniel K. Benjamin, in Fisheries are Classic Example of the "Tragedy of the Commons", cites research by Grafton, Squires, and Fox to support the idea that privatisation can solve the overfishing problem:

According to recent research on the British Columbia halibut fishery, where the commons has been at least partly privatized, substantial ecological and economic benefits have resulted. There is less damage to fish stocks, the fishing is safer, and fewer resources are needed to achieve a given harvest. Another possible solution, at least for some areas, is fishing quotas, so fishermen can only legally take a certain amount of fish.

A more radical possibility is declaring certain areas of the sea "no-go zones" and make fishing there strictly illegal, so the fish in that area have time to recover and repopulate. Controlling consumer behaviour and demand is a key in mitigating action. Worldwide a number of initiatives emerged to provide consumers with information regarding the conservation status of the seafood available to them. The Guide to Good Fish Guides lists a number of these.

Marine Stewardship Council

The Marine Stewardship Council (MSC) is an independent, global, non-profit organisation which was set up in 1997 to find a solution to the problem of overfishing. It has developed an environmental standard for sustainable and well-managed fisheries. Environmentally responsible fisheries management and practices are rewarded with the use of its blue product ecolabel.

Consumers concerned about overfishing and its consequences are increasingly able to choose seafood products which have been independently assessed against the MSC's environmental standard and labelled to prove it. This enables consumers to play a part in reversing the decline of fish stocks. As of January 2007, 22 fisheries around the world have been independently assessed and certified as meeting the MSC standard, and there are nearly 500 seafood products sold by retailers in 25 countries around the world.

Their 'where to buy' page lists all currently available certified seafood. Fish and Kids is an MSC project to teach schoolchildren about marine environmental issues, including overfishing.

GROWTH OF THE SEAFOOD PROCESSING SECTOR

The seafood processing industry in India is spread all along the maritime states of the country. The total export has touched 612641 mt valued at ₹. 83640 million (US $ 1853 million) during 2006-07. Gujarat is a major seafood exporting state and the export during 2006-07 from the state was 188166 mt worth ₹. 12650 million (US $ 281 million). The state's share in the total exports was 30.71 percent in terms of quantity and 15.12 percent in terms of value. Twenty two of the total 64 processing units in Gujarat are EU approved.

THE WORKFORCE OF SEAFOOD

The quantum of work in the seafood industry is directly related to the availability of raw material and tends to be seasonal (Anon, 2001). In the present study it was observed that the peak period is from September to April and the lean season from June to August. The pre-processing work includes grading, sorting, distribution, evisceration, cutting, slicing and cleaning in case of fish, peeling, cleaning and grading in case of shrimp, evisceration, cleaning and grading in case of cephalopods and cleaning of the processing hall. The processing work involves grading, slicing in cephalopods, packing in trays and cartons, loading, freezing, and cold storage.

Distribution of Workforce by Gender

The distribution of workforce in the units study. The data clearly indicates that the participation of women is mostly confined to the floor level which is categorised as unskilled, but there is more drudgery. Their participation in other higher categories, where there is more responsibility or decision making involved, is negligible. At the floor level, the male female ratio in the processing sector is 1:1.74, with the ratio being higher for the contract or temporary category where for every man, two women are employed. In the regular or permanent category the ratio is 1:1.66. Further 53.13 percent of the respondents reported that their immediate seniors were male and 46.88 percent reported having female supervisors. This is also an indication of the gender differentiation as far as specific jobs are concerned, and more men than women are dominant in the supervisory categories. At the managerial level the participation of women was just 4 percent. Women are also seen in the quality control sector mainly as technologists.

PLANT LOCATION, PHYSICAL ENVIRONMENT AND INFRASTRUCTURE

Early considerations in building a new plant is the identification of a suitable location. A number of factors should be considered such as physical, geographical and infrastructure available. A plant must be located on a plot of adequate size (for present needs and future developments), with easy access by road, rail or water. An adequate supply of potable water and energy must be available throughout the year at a reasonable cost. Special considerations must be given to waste disposal. Seafood processing plants usually contain significant amounts of organic matter which must be removed before waste water is discharged into rivers or the sea. Also solid waste handling needs careful planning, and suitable space, away from the plant, must be allocated or be available. Assessment of pollution risk from adjacent areas must also be considered. Contaminants such as smoke, dust, ash, foul odours (*e.g.*, neighbouring fish meal plant using poor raw material) are obvious, but even bacteria may have to be considered as airborne contaminants (*e.g.*, proximity of a poultry rearing plant upwind may be a source of Salmonella sp).

The immediate physical surroundings of a seafood factory should be landscaped and present on attractive view to the visitor (-or potential buyer of products). However, this should be done in a way so rodents and birds are not attracted. Shrubbery should be at least 10 m away from buildings and a grassfree strip covered with a layer of gravel should follow the outer wall of buildings. This allows for thorough inspection of walls and control of rodents.

Buildings, Construction and Layout

A food processing plant shall provide:

- Adequate space for equipment, installations and storage of materials.
- Separation of operations that might contaminate food.
- Adequate lightning and ventilation.
- Protection against pests.

The requirements of external walls inclusive roofs, doors and windows are that they should be water-, insect- and rodent proof. Internal walls, on the other hand, should be smooth, flat, resistant to wear and corrosion, impervious, easily cleanable and white or light coloured. Also the floors should ideally be impervious to spillage of product, water and disinfectants, durable to impact, resistant to disinfectants and chemicals used, slip resistant, non-toxic, non-tainting and of good appearance and easy repairable. Floors should be provided with slope to drains to prevent formation of puddles. The technical requirements, choice of materials, cost, etc., to obtain these goals may be found in a number of publication such as Shapton and Shapton (1991), Imholte (1984), Troller (1983).

The general layout and arrangements of rooms within a processing establishment is important in order to minimize the risk of contamination of the final product. A large number of bacteria (pathogens and spoilage bacteria) enter with the raw material. To avoid cross contamination, it is therefore essential that raw material is received in a separate area and stored in a separate chillroom. From here the sequence of processing operations should be as direct as possible - and a "straight line" process flow is regarded as the most efficient.

This layout minimizes the risk of recontamination of a semi-processed product. A clear physical (*e.g.*, a wall) segregation between "clean" and "unclean" areas is of prime importance. "Unclean" areas are those where raw material is handled and often a cleaning operation (wash) or, *e.g.*, a heat treatment (cooking of shrimp) is marking the point, where the process flow goes from "unclean" to "clean" areas. Thus a "clean" area is defined by ICMSF (1988) as an area where any contaminant added to the product will carry over to the final product, *i.e.*, there is no subsequent processing step that will reduce or destroy contaminating microbes. Other terminologies used for "clean" areas are "High Care Areas".

Also cooled rooms must be separated from hot rooms where cooking, smoking, retorting, etc., are taking place. Dry rooms must be separated from wet rooms and ventilation must be sufficient to remove excess humidity. The separation between the clean and unclean areas must be complete. There should be no human traffic between these areas, and equipment and utensils used in the unclean areas should never be used in the clean area. This means that there should also be separate wash and hygiene facilities for equipment and personnel in these areas. For easy identification the personnel should wear different coloured protective clothing for different operations (*e.g.*, white in the clean area and blue in the unclean).

Equally important in layout and design of food factories is to ensure that there are no interruptions and no "dead ends" in the product flow, where semiprocessed material can accumulate and remain for a long time at ambient temperature. Time/temperature conditions for products during processing are extremely important critical control points (CCPs) in order to prevent bacterial growth. This means that a steady and uninterrupted flow of all products is necessary in order to have full control of this critical factor. If any delays in product flow are necessary, the products should be kept chilled.

In addition to facilitate product flow, the factory layout and practices should ensure that:

- All functions should proceed with a no of criss-crossing and backtracking.
- Visitors should move from clean to unclean areas.
- Ingredients should move from "dirty" to "clean" areas as they become incorporated into food products.
- Conditioned (*e.g.*, chilled) air and drainage should flow from "clean" to "dirty" areas.
- The flow of discarded outer packing material should not cross the flow of either: unwrapped ingredients or finished product.
- There is sufficient space for plant operations including processing, cleaning and maintenance. Space is also required for movement of materials and pedestrians
- Operations are separated as necessary. There are clear advantages in

minimizing the number of interior walls since this simplifies the movement of materials and employees, makes supervision easier, and reduces the area of wall that needs cleaning and maintenance (the list is partly after Shapton and Shapton 1991).

Utensils and Equipment

A great variety of utensils and equipment is used in the fish industry. There is an abundance of advice and regulations available concerning the requirements for equipment. All of them agree that the food equipment should be non-contaminating and easy to clean. However, the degree of stringency in hygienic requirements must be related to the product being processed. Raw fish for example, do not require the same standard of hygiene as cooked and peeled shrimp. Criteria for hygienic design are particularly important for equipment used in the later stages of processing and particularly after a bacteria-eliminating processing step.

There are seven basic principles for hygienic design agreed upon by a working party appointed by Food Manufacturers Federation (FMF) and Food Machinery Association FMA (FMF/FMA 1967) as quoted by Hayes (1985):

- All surfaces in contact with food must be inert to the food under the conditions of use and must not migrate to or be absorbed by the food.
- All surfaces in contact with food must be smooth and non-porous so that tiny particles of food, bacteria, or insect eggs are not caught in microscopic surface crevices and become difficult to dislodge, thus becoming a potential source of contamination.
- All surfaces in contact with the food must be visible for inspection or the equipment must be readily disassembled for inspection, or it must be demonstrated that routine cleaning procedures eliminate possibility of contamination from bacteria or insects.
- All surfaces in contact with food must be readily accessible for manual cleaning, or if not readily accessible, then readily disassembled for manual cleaning, or if clean-in-place techniques are used, it must be demonstrated that the results achieved without disassembly are the equivalent of those obtained with disassembly and manual cleaning.
- All interior surfaces in contact with food must be so arranged that the equipment is self emptying or self draining.
- Equipment must be so designed as to protect the contents from external contamination.
- The exterior or non-product contact surfaces should be arranged to prevent harbouring of soils, bacteria or pests in and on the equipment itself as well as in its contact with other equipment, floors, walls or hanging supports.

In the design and construction of equipment it is important to avoid dead areas where food can be trapped and bacterial growth take place. Also dead ends (*e.g.*, thermometer pockets, unused pipe work T-pieces) must be avoided, and any piece of equipment must be designed so the product flow is always following the "first in first out" principle.

Cleanability of equipment involves a number of factors such as construction materials, accessibility and design.

The most common design faults which cause poor cleanability are:

- Poor accessibility (- equipment should be placed at least 1 m from wall, ceiling or nearest equipment).
- Inadequately rounded corners (minimum radius should be 1 cm, but 2 cm is regarded as optimum by the American 3-A Sanitary Standards Committee.
- Sharp angles.
- Dead ends (including poorly designed seals).

One general problem of food processing involves the extremes of temperature, abundant use of water, condensation and contamination of food from overhead pipes and surfaces.

Equipment design must consider this and include proper protection. Equipment design is one of the major problems in modern food hygiene. A great number of new machines and equipment are designed and constructed without proper attention to the fact that these tools have to be cleaned and sanitised. The EEC Directive 89/392/EEC (EEC 1989) addresses machinery safety and hygiene regulations.

Some of the highlights are:

- Machinery containing materials intended to come in contact with food must be designed and constructed so these materials can be cleaned before each use.
- All surfaces and their joinings must be smooth, with no ridges or crevices that could harbour organic materials.
- Assemblies must be designed to minimize projections, edges and recesses. They should be constructed by welding or continuous bonding, with screws, screwheads and rivets used only where technically unavoidable.
- Contact surfaces must able to be readily cleaned and disinfected, and built with easily dismantled parts. Inside surfaces must be curved in a way to allow through cleaning.
- Liquid derived from foods, as well as cleaning, disinfecting and rinsing fluids should be able to be readily discharged from machinery.
- Machinery must be designed and constructed to prevent liquids or living creatures - primarily insects - from entering and accumulating in areas that cannot be cleaned.

- Machinery must be designed and constructed so that ancillary substances, such as lubricants, do not come in contact with food.

The directive also sets out a certification system where machinery is checked for compliance and tagged with an EC mark if found to be satisfactory. Certification is not retrospective and manufacturers have two years to bring new machinary into compliance. Apart from literature already cited, additional useful material and information on hygienic design are found in Anon. (1982, 1983), Milledge (1981) and Katsuyama and Strachan (1980).

PROCESSING PROCEDURES

Processing procedures are Critical Control Points (CCP-2) in the processing of all food products. All processing techniques and procedures therefore must be designed and aimed at management of contamination and/or growth of microorganisms in food. Such procedures are termed "Good Manufacturing Practices" (GMP).

Detailed codes for GMP must be elaborated for each factory and each processing line (like the HACCP-concept). However a number of details to be included in the GMP-codes have been elaborated by regulatory agencies and international organisations.

The most comprehensive example is the work undertaken by the "Codex Alimentarius Commission" of the United Nations, who has published a series of Recommended Codes of Practices (Codex Alimentarius 1969-) including general principles of food hygiene (Vol. A) and a number of fish products(Vol. B) including codes for fresh fish, canned fish, frozen fish, shrimp, molluscan shellfish, lobsters, crabs, smoked fish, salted fish and minced fish. These codes are continuously updated and should be consulted for detailed information on recommended processing procedures.

Personal Hygiene

Personal hygiene is a CCP-2 in preventing microbial contamination or any foreign body contamination of fish products. A list of 15 basic points related to personal hygiene has been drawn up by Thorpe (1992) and are shown below: Personal hygiene requirements for personnel working in production areas and materials warehouses:

- Protective clothing, footwear and headgear issued by the company must be worn and must be changed regularly. When considered appropriate by management, a fine hairnet must be worn in addition to the protective headgear provided. Hair clips and grips should not be worn. Visitors and contractors must comply with this regulation.
- Protective clothing must not be worn off the site and must be kept in good condition. If it is in poor condition, inform your supervisor immediately.

- Beards must be kept short and trimmed and a protective cover worn when considered appropriate by management.
- Nail varnish, false nails and make up must not be worn in production areas.
- False eyelashes, wrist watches and jewellery (except wedding rings, or the national equivalent, and sleeper earrings) must not be worn.
- Hands must be washed regularly and kept clean at all times.
- Personal items must not be taken into production areas unless carried in inside overall pockets (handbags, shopping bags must be left in the locker provided).
- Food and drink must not be taken into or consumed in areas other than the tea bars and the staff restaurant.
- Sweets and chewing gum must not be consumed in production areas.
- Smoking or taking snuff is forbidden in food production, warehouse and distribution areas where 'No Smoking' notices are displayed.
- Spitting is forbidden in all areas on the site.
- Superficial injuries (*e.g.*, cuts, grazes, boils, sores and skin infections) must be reported to the medical department or the first aider on duty via your supervisor and clearance obtained before entering production areas.
- Dressings must be waterproof and contain a metal strip as approved by the medical department
- Infectious diseases (including stomach disorders, diarrhoea, skin conditions and discharge from eyes, nose or ears) must be reported to the medical department or first aider on duty via your supervisor. This also applies to staff returning from foreign travel where there has been a risk of infection.
- All staff must report to medical department when returning from both certified and uncertified sickness.

CLEANING AND DISINFECTION IN PROCESSING PLANT

Water Quality in Processing and Cleaning

As a general rule, water used for all purposes in food production must meet drinking water standards. It is noted that a universal list of biological and physico-chemical parameters for drinking water does not exist. The WHO Guidelines for drinking water quality and the guidelines prepared by EU (WHO, 1984; EEC, 1980) are similar with regard to microbiological contamination. The same situation applies concerning state regulations and only physico-chemical requirements for drinking water differ in particular countries. Disinfectant residues should be monitored where possible and the bacteriological quality periodically checked. Turbidity, colour, taste and odour are also easily monitored parameters. If there are local problems with chemical constituents (fluoride, iron) or contaminants from industry or agriculture (*e.g.*, nitrate, pesticides, mining wastes) these should (hopefully) be monitored and dealt with by the water suppliers.

Very often water must undergo treatment disinfection prior to use. The following chemicals are used as disinfectants: chlorine, chloramine, ozone or UV irradiation. Chlorination is the cheapest form of treatment and monitoring of chlorine is relatively easy. According to WHO (1984) the concentration of chlorine in water should be in the range 0.2-0.5 mg/l. For sanitation purposes it may reach 200 mg/l, but in order to avoid corrosion lower concentrations are advised (50-100 mg/l).

Cleaning and Disinfection

Cleaning and disinfection are the most frequent operations in modern food processing. Carelessness may cause considerable economic loss, and loss of reputation on the market. The hygienic standards respected in processing plants depend on kinds of production. For example, in the cannery they will be more strict than in plants where fish is only gutted and stored in ice and its shelf life is rather short. Regarding all other technological operations and processes, cleaning and disinfection procedures must follow detailed instructions and responsible personnel be assigned.

Various steps should be included in a complete cycle of cleaning and disinfection:

- Remove food products, clear area from bins, containers, etc.
- Dismantle equipment to expose surfaces to be cleaned. Remove small equipment, parts and fittings to be cleaned in a specified area. Cover sensitive installations to protect them against water, etc.
- Clear the area, machines and equipment of food residues by flushing with water (cold or hot) and by using brushes, brooms, etc.
- Apply the cleaning agent and use mechanical energy (*e.g.*, pressure and brushes) as required.
- Rinse thoroughly with water to completely remove the cleaning agent after the appropriate contact time (residues may completely inhibit the effect of disinfection).
- Control of cleaning.
- Sterilisation by chemical disinfection or heat.
- Rinse off the sterilant with water after the appropriate contact time. This final rinse is not needed for sterilants, *e.g.*, H_2O_2 based formulations which decompose rapidly.
- After final rinsing, equipment is reassembled and allowed to dry.
- Control of cleaning and disinfection.

- In some cases it will be good practice to re-disinfect (*e.g.*, with hot water or low levels of chlorine) just before production recommences.

As mentioned above, only agents and disinfectants permitted by adequate regulations, can be used for cleaning and disinfection operations. During their use precautionary measures must be observed and this requires proper training of personnel.

QUALITY ASPECTS OF FRESHWATER FISH PROCESSING

Public Health Aspects

The term quality has many different implications, *e.g.*, product excellence, value, nutrition, safety for consumer, etc. This section discusses quality requirements with respect to safety for the consumer and quality control principles.

In a free market economy the producers are responsible for food quality and they are controlled by the competent authorities according to approved procedures. Certain countries or groups of countries, *e.g.*, European Union, formulate regulations specifying requirements concerning health quality, wholesomeness of raw materials and food/fish products and concerning permissible limits for chemical contaminants (heavy metals, PCBs, etc.) or biological infestants (parasites, microbes, etc.). Other regulations concern quality of water provided for food processing.

These regulations are of rather general character but there are others which concern health conditions for processing and placing of products on the market. Due to an almost complete lack of detailed standards for individual products, the regulations on labelling are of great importance, especially if the "fair trade" principle and consumer interest are to be taken into account. All these groups of obligatory regulations should ensure production of food which is safe for the consumer.

Additionally the monitoring of raw materials is a complementary part of activities carried out according to requirements contained in regulations. It provides the competent authorities, responsible for supervision of production, with information about potential hazard.

As mentioned earlier, producers are responsible for food quality. Besides the competent authorities such as the Ministry of Health, Ministry of Agriculture and Veterinary Services and consumer organisations or associations, producers also participate in creating new food laws. Such cooperation enables rules corresponding to industrial reality to be created which at the same time ensure consumer safety. Moreover, guides such as: Good Manufacturing Practice, prepared by producer associations, or Codes of Good Manufacturing Practice elaborated by FAO/WHO, are complementary tools widely used in assuring product quality. They lay down detailed technological procedures and recommendations for production. Provided they are respected by the producer the expected product quality is reached and consumer safety ensured.

Familiarisation with principles contained in these guides and codes is important especially in the case of small food processing plants which unfortunately are often directed by people without adequate professional qualifications and training. Rules and regulations of US FDA codes and standards of FAO/WHO, Council Directives of EEC (EU) set out an approach to the issue of health quality assurance. According to the above regulations, the main principle is that fish and fish products constitute a source of potential health hazard and danger for consumer safety. Many requirements, regulations, supervision, controls, inspections and governmental interventions stem from that principle. Listed below are some basic regulations on requirements for fish and fish products from aquaculture, which are either compulsory or have been introduced in European Union countries. These documents concern also Third Countries which export fish and fish products to the EU market.

Requirements in this context are covered by the following documents:

- Council Directive (91/67/EEC) of 28 January 1991 concerning the health conditions of animals destined for marketing and originating from aquaculture
- Council Directive (91/493/EEC) of 22 July 1991 laying down the health conditions for the production and the placing of fishery products on the market
- Council Regulation (EEC No 3759/92) of 17 December 1992 on the common organisation of the market in fishery and aquaculture products

This first directive states that animals must:

- Be free of clinical signs of disease on the day of loading;
- Not be directed for processing in order to liquidate such diseases as:
- Infectious haematopoietic necrosis (IHN),
- Viral haemorrhagic septicaemia (VHS),
- Infectious pancreatic necrosis (IPN),
- Bacterial kidney disease (BKD),
- Spring viremia of carp (SVC),
- Enteric red mouth disease (ERM)
- Gyrodactylosis (Gyrodactylus salaris)
- Myxobolosis (Myxosomiasis-whirling disease);
- Not come from a farm which is closed due to diseases, and must not be in contact with fish from such a farm;
- Be subject to the same requirements if directed for farming;
- Be delivered, in the case of aquaculture fish, in the shortest possible time to the destination, and the change of water must only be done in specified places. Such places must be known to European Union countries;

- The Commission checks if regions are free of diseases, approves them and can, within reason, also revoke approval of the decision; and finally makes a list of approved fish farms;
- Permission may be granted for placing on the market fish from aquaculture and from regions not approved but under special conditions. Such instances require documentation confirming the wholesomeness of fish from this region, and this must be issued by official inspectors, for example, by the competent veterinary authorities.

The region can be approved as free of diseases if it meets at least two requirements:

1. The diseases listed above did not occur for at least four years,
2. All fish farms located in this region are under continuous veterinary supervision and are inspected at least twice a year.

Veterinary inspection should cover the visual assessment of aquaculture fish wholesomeness, taking of fish samples and immediately sending them to the competent laboratory. Each farm must record all necessary data pertaining to the wholesomeness of fish including the official certificate of laboratory analysis. It is pointed out that only certificates issued by official control authorities are valid and placing of fish in the fish farm or for sale has to be formally documented. The above rules concern all the fish being sold domestically and not only the fish directed for European markets. Fish processing plants which, apart from farming carry out processing must obtain the approval of veterinary authorities to export their products. Acquiring a registration number is a formal approval to export fish. This registration number enables identification of the fish product, and this number must be shown on the label of each package and on the relevant documents. Aquaculture fish and fish products exported to the EU must fulfil the conditions specified in Council Directive 91/493/EEC. The level of requirements in this Directive indicates that many fish processing plants will face great difficulty in obtaining an export licence.

Thus each establishment should draw up its own production programme covering for example:

- Kind of production (for example fresh fish, frozen fish, canned products, etc.),
- Volume of production (daily, annual),
- Production rooms and store rooms,
- Technical equipment,
- Sanitary facilities for staff,
- Written schedule of quality assurance system and adherence to it

The competent veterinary authorities evaluate the production capacity and possibilities of fulfilling the production programme, taking into account the technical abilities and insurance of adequate sanitary conditions.

Quality Control

Control is traditionally limited to control of the final product. Practice has proved that this is not sufficient and that quality control should be carried out during all stages of production, starting from a contract on supply of raw material, through all the phases of processing, to storage and distribution of final products. Such an approach is not quality control but constitutes quality assurance, which covers the entire production chain. Below, principles relating to quality control are presented with regard to the main individual operations and procedures in fish processing.

Drawing up a Contract for Raw Material Supply

The contract for supply of raw material should cover all specific requirements, for example: size of fish, closed seasons, level of chemical contaminants in fish and in the water from which the fish comes, and chemical measurements should be made by an institution which deals with monitoring of environment. Sometimes, especially in the case of export, the buyer/customer may have additional demands, *e.g.*, an indication of the level of chemical contaminants other than standard ones. The buyer should ensure that he will receive the health certificate for his raw material and that the certificate was issued by the official control authorities. He should also obtain confirmation that the fish was stored properly prior to sale (for example, that fish was iced with a proper amount of ice and that the quality of the ice was satisfactory; that it was stored in cool store rooms and that it was transported by appropriate means). The contract may specify that some of these demands be passed to the receiver.

Receiving and Storage of Raw Material

This control step determines the quality of the final product and should be carried out extremely carefully.

In general it consists of three elements:

- Temperature control of fish during transportation (temperature record)
- Temperature control of fish and control of icing
- Quality control of purchased fish

The temperature of the fish hold in the transport system is usually registered automatically or periodically by a driver. This temperature record is part of the documentation on fish shipment. Measurements of real temperature of fish tissue and control of icing are made on random samples. Apart from these elements the cleanliness of the means of transport and the containers, and the labelling, are checked. The number of samples/packages with fish to be further assessed depends on lot size, and it should be specified clearly in compulsory procedures or codes of good manufacturing practice, perhaps in the standards or contract specifications. The temperature of purchased fish should be close to ice melting temperature and not higher than 4° C. The samples of fish taken for temperature measurement are at the same time the samples examined for quality control of raw material. Usually in the case of medium size batches eight packages are taken and in each package three temperature checks are made.

Detailed quality assessment is made according to requirements laid down in procedures, codes or standards if the latter exist. Such an assessment is carried out on an average sample from a set of randomly selected packages.

The sensory analysis of raw material is a main part of control, and it allows full characteristics of the fish investigated to be obtained. This analysis includes appearance of skin, eyes, gills and fish as a whole, colour of fish tissue; damage to fish, springiness of meat tissue, flavour of individual organs; flavour, taste and texture of meat tissue after cooking. Occurrence of inadmissible features like for example sour smell of gills, strange/unfamiliar smell of meat or fish as a whole causes that raw material is disqualified and excluded as a material intended for processing. In the case of live fish their appearance and movement in the water in a container are assessed.

The kind and the degree of infestation with parasites determines further procedure. If the presence of parasites which are harmful for humans is detected, fish cannot be sold as fresh. As mentioned above, this matter should be considered by the receiver when the contract is prepared. The final result of quality control of raw material is decisive with respect to further procedure during fish processing. Generally when fish is qualified as conforming with requirements and cooled properly it is placed in cold stores or transported direct to the processing line. Ice is added to fish cooled insufficiently and this is placed in cold store. The temperature inside the cold store should be close to 0° C and should be continuously recorded. If temperature cannot be registered automatically, measurements should be taken not less frequently than every two hours.

Quality Control During the Production Process

The quality control programme during the production process depends on the profile of production carried out in the processing plant. Each processing plant must draw up a flow chart of the entire process starting from the raw material through every individual operation and process to the final product and with all quality control points indicated. Criteria for selection of control points depend on potential hazards which, in the case of lack of proper handling, can cause a risk for both the food and the consumer.

For example, control of temperature during individual operations, their duration, concentration of food additives, etc., are typical and critical parameters measured at control points. Technological supervision is responsible for use of adequate processing parameters. Quality control personnel are responsible for monitoring these parameters and in the case of deviation they should undertake proper corrective action. The final step in production control is the quality control of the final product according to technical requirements and specifications included in the contract or standards if the latter are compulsory. Such assessment is carried out according to approved procedures with special regard to health quality requirements pertaining in a given country. This type of control will disappear in the future because an introduction of quality assurance systems, as a continuous control throughout the entire processing procedure, will eliminate this traditional form of control.

Quality control personnel are also responsible for supervision of assurance of cleanliness and disinfection of production lines and processing rooms. Maintenance of cleanliness and disinfection should be carried out in accordance with a programme approved by the local veterinary service.

The quality control staff assures adherence to this programme which especially concerns:

- Types of detergents/disinfectants and concentrations used;
- Compliance with procedures of cleaning/washing and disinfection;
- Arrangement of periodic microbiological measurements on the surface of equipment and processing machines;
- Control of personal hygiene of staff including working clothes and sanitary fittings in the plant.

In summary, the quality control staff is responsible for carrying out this programme and for the sanitary-hygienic conditions of the processing plant and for maintaining the documentation relating to these activities.

Storage and Distribution of Freshwater Fish Products

The fish products directed for the storage or for the purchaser are random checked by the internal quality control staff.

This control, *inter alia*, concerns:

- Proper packaging materials and labelling (according to official requirements);
- Duration and temperature of storage;
- Proper conditions of storage, for example adequate ice, temperature etc;
- Choice of means of transportation and hygienic conditions (cleanliness, temperature record, etc.);
- Proper loading (*e.g.*, arrangement of load in vehicles).

OVERVIEW OF THE SEAFOOD INDUSTRY

The world seafood industry plays a significant role in the economic and social wellbeing of nations, as well as in the feeding of a significant part of the world's population. Fishing and fish farming has emerged as one of the major food processing occupations of mankind. In ancient times, economically and socially backward people were employed in this profession. The advent of modern mechanised fishing vessels has brought vast changes in the attitude of the public fishing and seafood processing. From low income and socially backward communities the profession has shifted to the hands of industrialists and technologists. Today fishing and processing activities provide employment to millions of people around the world.

SOURCES OF FISH AND SHELLFISH CONSUMED IN THE U.S.

The commercial seafood products consumed by Americans at home or in restaurants or other foodservice establishments primarily come from three different sources: U.S., commercial fisheries, U.S., aquaculture production, or imports brought into the U.S., from other countries. Seafood is also a unique food in that a large amount of fish and shellfish are harvested from the wild by individuals for recreational purposes and some of that harvest is also consumed. The following information provides an overview of the types of fish and shellfish that comes from these four different sources.

U.S., Commercial Fisheries

Commercial fishery landings of edible fish and shellfish were 6.5 billion pounds in 2010. Over 80 per cent of the total commercial catch was finfish, but shellfish represented more than 50 per cent of the total value. The major fish and shellfish species harvested by U.S., fishermen ranked by both volume and value are provided in the tables on this page. Over 50 per cent of all U.S., landings were fish caught by trawlers in the Pacific Ocean including groundfish like Pacific cod, flounders, hake, ocean perch, Alaska pollock, and rockfishes. Other important commercial Pacific Ocean species are salmon, halibut, Dungeness, King and Snow crab, tuna, and squid.

In the Atlantic Ocean, some of the most economically important species include: scallops, lobster, clams, blue crab, oysters, and herring. Shrimp is an important fishery in the Gulf of Mexico and South Atlantic. Another important fishery is groundfish species caught by trawlers in the North Atlantic from Chesapeake Bay through New England that include: butterfish, Atlantic cod, cusk, haddock, hake, ocean perch, and Atlantic pollock. A variety of finfish species such as tuna, flounder, grouper, snapper and other reef fish are important fisheries in the South Atlantic and Gulf of Mexico. The menhaden fishery in the Mid-Atlantic and Gulf of Mexico is also important, but is not used for human food but for bait or conversion to fish oil and fish meal that is used in a variety of products.

U.S. Aquaculture Production

The production of farm raised fish and shellfish in 2009 was about 725 million pounds worth $1.2 billion. In the United States, the amount of fish and shellfish harvested from the wild annually is about 8 times greater than the amount produced by domestic aquaculture farms. Pond raised catfish represents about two thirds of the total farm raised seafood products produced annually in the U.S. Other important domestically produced aquaculture food products in order of the quantity produced include: crawfish, salmon, trout, oysters, tilapia, striped bass, clams, shrimp, and mussels.

Imported Seafood Products

China is the largest producer of seafood products in the world, and Japan and the U.S., are the largest importers of seafood products in the world. Over three fourths of the seafood consumed in the U.S., is imported from other countries. In 2009, 5.5 billion pounds of edible fishery products valued at $14.8 billion were imported into the U.S., Shrimp is the most important imported seafood product, and over 1.2 billion pounds of shrimp were imported in 2010.

Thailand was the leading U.S., supplier of shrimp followed by Ecuador, Indonesia, China, Vietnam and Mexico. Tuna was the second most important imported product in 2010, and an almost equal amount of canned tuna and fresh and frozen tuna were imported that year. Major suppliers of canned tuna are Thailand, Philippines, Indonesia, Vietnam and Ecuador. Freshwater fish fillets ranked third in volume for all seafood products imported into the U.S., in 2010.

A major part of this product category are the Vietnamese fish species, called pangasius, basa or swai in U.S., markets. Other important products in order by volume imported include: salmon from Norway, Canada, and Chile; groundfish species like cod, haddock, pollock and hake from Canada and Northern Europe, crabs and crabmeat from Southeast Asia, and frozen fish blocks used to make fish portions and sticks from China, Russia, Canada, and Iceland.

Recreational Fisheries in the U.S.

U.S. Fish and Wildlife Service estimates from 2006 indicate that about 30 million people in the U.S., were engaged in recreational fishing that year. About three fourths of these anglers fished in freshwater including the Great Lakes, other lakes, rivers, streams and ponds. About one fourth fished in saltwater either in the ocean or in near coastal areas. In 2010, the National Marine Fisheries Service estimated that the total weight of the harvested saltwater catch was 212 million pounds, but no estimate of the weight of freshwater fish was available.

The most frequently caught fish in lakes, rivers, streams and ponds were: black bass, panfish, catfish/bullhead, trout, crappie, white bass, striped bass, and striped bass hybrids. In the Great Lakes the most frequently caught fish were: walleye, sauger, perch, salmon, lake trout, black bass, and steelhead.

The most frequently caught marine saltwater fish in the U.S., included: flatfish such as flounders or halibut, red drum, sea trout, striped bass, bluefish, and salmon. There are many local and regional differences in the type of saltwater species caught in the three major saltwater areas of the country that include the Atlantic Ocean, Gulf of Mexico and Pacific Ocean.

Seafood Processing

The processing sector of the seafood industry converts the whole fish or shellfish harvested by fishermen or produced by aquaculture operations in the U.S., or in other countries into the products that are sold at retail stores or restaurants. The National Marine Fisheries Service estimated that in 2010 the value of edible processed seafood products in the U.S., was $8.5 billion. The annual U.S., production of raw (uncooked) fish fillets and steaks, including blocks, is around 500 million pounds, and the major species processed were Alaskan Pollock, salmon, cod, hake, flounders and haddock.

The combined production of fish sticks and portions has been between 200 and 300 million pounds over the past decade, and the production of breaded shrimp between 75 and 150 million pounds. The pack of canned fishery products varies from year to year between 500 million and a billion pounds with tuna, salmon and clams being the major canned products produced in the U.S.

Primary processors generally convert whole fish into fish fillets, steaks or loins or shuck or cook raw shellfish or remove the edible meat. These edible portions are then packed in some way and distributed as fresh refrigerated products or are frozen prior to distribution to wholesalers or directly to retail stores or restaurants. Other processors pack these edible portions into cans or other containers and apply a heat process to eliminate microorganisms that could cause the product to spoil or cause foodborne illness.

Canned products are treated to sterilise their contents and can be stored without refrigeration. Pasteurised products are heat treated in a way that eliminates most but not all microorganisms and must be stored under refrigeration. Other processes could include the use of high pressure, irradiation or other treatments to sterilise or pasteurise the seafood product.

Secondary processorsconvert fresh or frozen fish and shellfish products and other ingredients into the final products that are available in retail stores and restaurants. Examples of value added finished seafood products could include: smoked seafood products, sushi, seafood salads and sandwiches, and seafood entrees or meals.

Seafood Wholesale and Distribution

There is a large network of wholesale and distribution businesses in the U.S., that purchase seafood products from a variety of different sources, store them, assemble the items into orders for customers, and deliver them. There are many variations to this basic business model. Some businesses specialise in specific types of products or products from a specific geographic area.

Other businesses called "broadline" distributors buy and sell a full line of all types of products to meet their customers' needs. Other businesses may focus on the unique needs of specific customers such as retail stores, restaurants, or institutional buyers with the the military, prisons, schools or hospitals. This much needed commercial business network is responsible for sourcing, purchasing, transporting, storing and delivering the seafood products available in our Nation's diverse markets.

Seafood Retail Stores

Fresh, frozen and processed seafood products are primarily available to consumers for home consumption from retail stores. It has been estimated that about one third of the seafood consumed in the U.S., is purchased at retail stores for home consumption. There are many different types of retail stores with different business strategies.

Small independent stores often specialise in products caught by local fisherman but also supply popular items such as shrimp which may come from Asia. Large retail chains also offer a variety of products which may also include locally caught items and a variety of other products from other regions of the U.S., or the world. Although there is some variation in the availability of seafood across the U.S., in most areas there is a wide variety of choices for retail purchases of seafood products.

Restaurants and Foodservice

U.S., consumers spend about two thirds of their annual expenditures on seafood in restaurants, cafeterias or other types of foodservice businesses. Seafood is an important item on the menu for most foodservice operations and the selection varies depending on the type of consumer that is targeted and menu prices.

Fish portions, breaded shrimp, clams and other items are served in a variety of chain restaurants because of their consistent cost and good value. Portion sizes are easy to control and they can be prepared quickly and consistently. Seafood is frequently used as an ingredient in pasta or rice dishes and in sandwiches, wraps, soups and other entrees in a variety of mid-priced restaurants. Fresh seafood is also widely available in restaurants that feature one or more chefs that use their skills to creatively prepare different local or specialty items depending on availability and demand.

GLOBAL SUPPLY AND DEMAND

The world's population is expected to increase by 36 per cent in the years 2000 to 2030, from approximately 6.1 billion people to 8.3 billion. It is also expected that the estimated total seafood demand will be 183 million tones by 2030, but the estimated supply will be only 150 to 160 million tones. Thus, there is a sizable gap between demand and supply. However, global capture fisheries will be able to provide only 80-100 million tones of fish annually on a sustainable basis. The global seafood market is estimated at US$ 100 billion per annum. Also, the world demand for seafood increases by 3 per cent each year. The world largest seafood consumption in the world is by Japan, followed by European Union. The top five consumed species are salmon, shrimp, tilapia, catfish and crab (major consumption in China and India).

SEAFOOD PRODUCTION PROCESS

Products

There are several product categories from seafood industry, based on raw material type (fresh/frozen) and value-addition (degree of processing and value content). The following figure depicts few seafood product categories in a seafood industry.

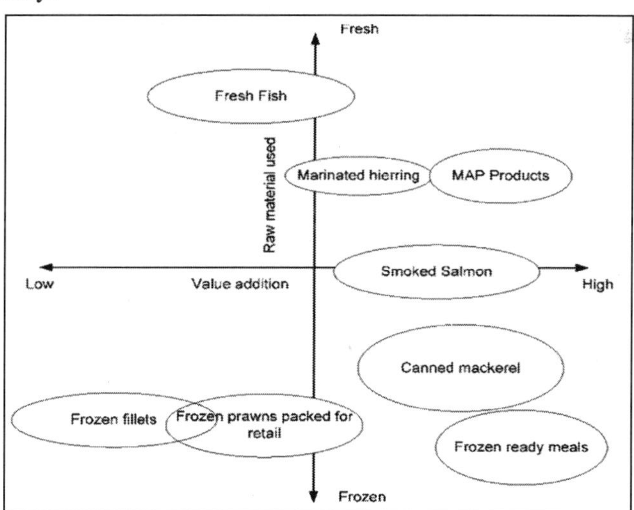

Fig. Seafood Products in a seafood industry

Important canned fish products are tuna packed as solid pack, chunks, flakes, grated or shredded in water or oil, sardines or sardine-like fishes in oil, tomato sauce or other types of sauce, pre-smoked sardines in oil or tomato sauce, kippers (pre-smoked herring), salmon, mackerel, fish paste products and pet food.

Table. Source and Approximate Yields of by-products from Various Fish Canning Operations

By-product	By-product yield from canning operations.		
	Tuna (%)	Sardine (%)	Salmon (%)
Pet food	4-6	-	-
Fish meal	30-35	20-30	30-35
Industrial oil	<5	5	-

Raw Material

Fish and Other Marine species

Many types of fish and other marine species are suitable for seafood production and the size of the individual fish varies from that of the smallest sardines to that of the largest tuna species. For some species like tuna and sardines canning is the most common processing method.

Other species, suitable for canning are salmon, mackerel, herring, clams, oysters, shrimps, octopus, crab and white fish paste products. To plan the handling and processing of seafood and to manage problems connected with all operations from transport to processing through storage, it is essential to know the properties of the species involved.

Table. Weight Percentage of Parts of the Common Fish Species Used in Canning

Species	Percentage of total weight					
	Head	Skin and flesh	Bones	Fins	Viscera	Ton/m^3
Atlantic herring	12.5	62.2	6.5	1.5	15.0	0.91
Sardines	21.0	58.0	6.5	2.5	9.5	0.85
Atlantic mackerel	22.5	52.0	8.0	1.0	19.5	0.96
Tuna	18.0	64.0	8.0	2.0	8.0	
Pink salmon	16.0	71.0	-	5.0	8.0	0.95

Ingredients

There is wide variety of liquid solutions available that can be added in seafood canning process.

Some of them are as listed below:

- Salt
- Olive oil
- Soya bean oil
- Tomato sauce

CLEANING AND SANITATION IN SEAFOOD PROCESSING

For both whole and cut fish, ozone sanitation is becoming the standard among progressive businesses in the industry because it offers superior pathogen control in cold water combined with no negative effect on the organoleptic qualities (taste, texture, smell and colour) of the product. Nothing in food processing is as perishable as the fresh smell and look of seafood. Seafood processing requires continuous sanitation of the plant, equipment, and surfaces, as well as the seafood products themselves. Poor or inadequate sanitation practices can contribute to reduced quality, increased spoilage and the potential of foodborne diseases. Ozone disinfection can be applied at multiple points in seafood processing:

Ozone can be utilised as an aqueous spray in seafood processing to minimize microorganism contamination on the seafood products themselves as well as the equipment and environment on the processing floor. Aqueous ozone can be sprayed directly on whole fish, after the skinner operation, post boner and fillet operation and continue to be used through packaging and further processing. This same ozone spray can be utilised before, during and after processing as a surface disinfectant to keep equipment and the environment sanitised. Ozone is an approved food additive, so it can be simultaneously applied at any time during processing, on the product and on the equipment.

Ozone Kills Biofilm

When sprayed on conveyors and automated cutting equipment during processing, ozone can eliminate biofilm and significantly reduce fat, oil and grease on all surfaces. This can reduce downtime for cleaning as well as provide further protection from cross-contamination in the processing line. It can be used as an additional point of intervention in the HACCP plan or it can be used in lieu of one or more traditional chemical or thermal processes.

Ozone Advantages

Ozone compares favourably with traditional disinfectants used in seafood processing. It has a broader spectrum of efficacy than chlorine, peroxyacetic acid, acidified sodium chlorite, hydrogen peroxide and quaternary ammonia. Ozone is an extremely effective antimicrobial that kills all known pathogens including E. coli, Listeria, Salmonella, Campylobacter, Bacillus and Norovirus.

Unlike other disinfectants, ozone will penetrate and destroy biofilm. Biofilm is endemic to cutting surfaces and processing room surfaces, and must be eliminated for reliable product safety. Ozone leaves no harmful byproducts and requires no rinsing.

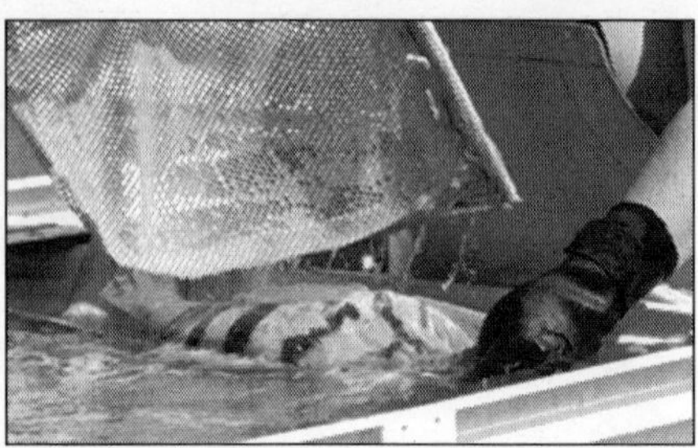

Applications in Seafood Processing

Ozone can be adapted to any existing aqueous or gaseous configuration with minimal retrofit of water and power. Aqueous ozone can be utilised through sprayers, showers, or cascades and can be plumbed into a flume. Gaseous ozone can be applied into any confined storage location.

WATER QUALITY IN PROCESSING AND CLEANING PROCEDURES

Water used for food processing is one of the important critical control points. This is true for water used as an ingredient, for water used as final rinse when cleaning equipment or water which is in any way likely to come into contact with the product. Most often it is just stated that the water should meet drinking water standards and both supply and quality are mostly taken for granted.

However, local standards may vary somewhat or may even be absent. The quality of the source water differs enormously from place to place as does the water treatment. The control exerted by the local regulatory authorities may also differ greatly depending on the local situation. Lastly, in-plant problems may sometimes render potable water unfit as drinking water at the final point of use.

So how can acceptable drinking water quality be defined ? What is the rationale behind these guidelines ? And what can the food processors do ? A universally accepted list of standards for biological and physico-chemical parameters for drinking water does not exist. WHO issued an excellent book called "Guidelines for drinking water quality", Vol 1, 2, and 3 (WHO 1984b). Volume 1 deals with the guideline values, Volume 2 contains monographs on each contaminant, and Volume 3 gives information on how to handle the water supplies in small, rural communities.

In this book, WHO recognises that very stringent standards cannot be used universally as this may severely limit the availability of water and instead, a range of guideline values for more than 60 parameters have been elaborated. A general review of the standards employed by WHO, EEC, Canada and USA is given by Premazzi *et al.* It is recognised that, *e.g.*, most of the rural wells all over the world would have difficulties meeting all the guideline values suggested. It goes without saying that all the parameters cannot be monitored so selection and priorities must be made based on hazard analysis and feasibility. Most nations (or in some cases even individual provinces) have their own guidelines or standards. The basic microbiological guideline values, however, do not differ so much from place to place.

Table. Microbiological Criteria (Guidelines) for Drinking Water Quality (WHO 1984b).

Organism in 100 ml[1)]	Guideline value	Remarks
Piped water supplies		
Treated water entering the distribution system		
fecal coliforms coliform chlorine,	0	turbidity < 1 NTU; for disinfection with
		pH preferably < 8.0, free chlorine residual 0.2-0.5 mg/1 following 30 min (minimum) contact
organisms	0	
Water in the distribution system		
fecal coliforms	0	
coliform the		in 95 per cent of samples examined throughout
		year -
organisms	0	in the case of large supplies when sufficient samples are examined.
coliform organisms	3	in an occasional sample but not in consecutive samples

Note: [1)] Multiple tube techniques (MPN procedure) and the membrane filtration technique havebeen considered as capable of yielding comparable information.

Table. Microbiological Criteria (Guidelines) for Drinking Water Quality (EEC 1980).

Parameters	Results: volume of of the sample (ml)	Guide level (GL)	Maximum admissible concentration (MAC) Membrane filter method	Multiple tube method (MPN)
Total coliforms	100	-	0	MPN<1
Fecal coliforms	100	-	0	MPN<1
Fecal streptococci	100	-	0	MPN<1
Sulphite-reducing clostridia	20	-	0	MPN<1
Total bacteria counts	1[1)]	10[1)]		
for water supplied for human consumption	1[2)]	100[2)]		

Note: [1)] Incubation at 37°C

[2)] Incubation at 22°C

In the case of water used for food production, it is of vital importance that these microbiological guideline values should be met since potentially pathogenic bacteria are capable of multiplying rapidly if they are introduced into foodstuffs making even initially low and non-infectious doses of bacterial pathogens, a hazard. Disinfectant residuals should be monitored where possible and periodic verifications of the bacteriological quality should be conducted. Turbidity, colour, taste and odour are also easily monitored parameters. If there are local problems with chemical constituents (*e.g.*, fluoride, iron) or contaminants from industry or agriculture (*e.g.*, nitrate, pesticides, mining waste) these should hopefully be monitored and dealt with by the water suppliers.

Effect of Water Treatment Including Disinfection on Microbiological Agents

Water treatments vary from region to region depending on the water sources available. While groundwater from sedimentary aquifers has undergone extensive filtration the water from hard rock aquifers or surface water sources should be filtered as part of the water treatment in order to decrease the content of particulates, microorganisms and organic and inorganic matter.

Parasites are removed to a large extent by filtration. The levels of bacteria and virus also decrease markedly and the removal mechanisms are both filtration and adsorption. The cation concentration influences adsorption, *i.e.*, increasing concentrations give rise to increased adsorption. Ca2+ and Mg2+ seem to be especially efficient. These small cations will decrease the repulsive forces between the soil particles and the microorganisms. Iron oxides also have a high affinity for viruses as well as bacteria. Ferric hydroxide impregnated lignite has even been suggested as a local filtration/adsorption media.

The disinfection efficiency is greatly affected by type of disinfectant, type and state of microorganism, water quality parameters such as turbidity (or suspended solids), organic matter, some inorganic compounds, pH and temperature. The "hardness" of the water may indirectly influence disinfection since deposits may harbour microorganisms and protect them from cleaning agents and disinfectants.

Type of Disinfectant

By far the most widespread disinfectant is chlorine but also chloramines, chlorine dioxide, ozone and UV light are being used in some instances. Chlorine is cheap and available in most places and monitoring free residual levels is simple. It is desirable to maintain a free residual chlorine level of 0.2-0.5 mg/l in the distribution system (WHO 1984b). For sanitation of clean equipment, up to 200 mg/l is used. To avoid corrosion lower concentrations of 50-100 mg/l and longer contact times (10-20 minutes) are often used. Chloramines are more stable but less bacteriocidal and much less efficient towards parasites and virus than chlorine. Chlorine dioxide is, if anything, more microbicidal than chlorine, especially at high pH, but there is concern with regards to the by-products. In the case of ozone and UV light there is no residual to monitor. Ozone seems to be very efficient towards protozoa. The efficiency of UV disinfection decreases markedly if there is any turbidity or dispersed organic matter and problems are often encountered due to lack of lamp maintenance.

Type and State of Microorganism

In the case of most disinfectants, the order of sensitivity is: vegetative bacteria > viruses > bacterial spores, acid-fast bacteria and protozoan cysts The sensitivity varies within groups and even within species. Our indicator bacteria are unfortunately among the more sensitive microorganisms and the presence of, *e.g.*, fecal coliforms in treated, disinfected water is therefore a very clear indication that the water contains potentially pathogenic microorganisms while the absence of such indicator bacteria do not guarantee pathogen-free water.

Bacteria from nutrient-poor media as well as otherwise stressed bacteria may also exhibit greatly increased resistance. Some of the effects mentioned on the efficiency of free chlorine.

Water Quality Factors

If microbes are associated with granular material or other surfaces the effect of a disinfectant such as chlorine decreases drastically. Attachment of Klebsiella pneumonia to glass surfaces may for example increase the resistance to free chlorine 150-fold. Organic matter may react and "consume" disinfectants such as chlorine and ozone and the presence will also interfere with UV light. The chloramines are less susceptible to organic matter.pH is important in disinfection with chlorine and chlorine dioxide with greater inactivation at low pH in the case of chlorine and greater inactivation at high pH in the case of chlorine dioxide. In general, higher temperatures result in increased inactivation rates.

Use of Non-Potable Water in a Plant

The use of non-potable water may be necessary for water conservation purposes or desirable because of cost. The water may, *e.g.*, be surface water, sea water or chlorinated water from can cooling. Relatively clean water such as chlorinated water from can cooling operations may be used for washing cans after closing before heat treatment, for transporting raw materials before processing (after the water has cooled off), for initial washing of boxes, for cooling of compressors, for use in fire protection lines in non-food areas and for fuming of waste material. It is absolutely necessary that potable and non-potable water should be in separate distribution systems which should be clearly identifiable. If potable water is used to supplement a non-potable supply the potable source must be protected against valve leaking, back-pressure, *e.g.*, by adequate air-gaps. Back-flow due to sudden pressure differentials or blockage of pipes have unfortunately occurred in many systems.

Table. Inactivation of microorganisms by free chlorine.

Organism	Water	Cl_2residues, mg/l	Temperature, °C	Ph	Time, min.	Reduction %	C*t1)
E. coli	BDF[2)]	0.2	25	7.0	15	99.997	ND[3)]
E. coli	CDF[4)]	1.5	4	?	60	99.9	2.5
E. coli + GAC[5)]	CDF	1.5	4	?	60	<<10	>>60
L. *pneumophila*	tap	0.25	20	7.7	58	99	15
(water grown)							
L. *pneumophila*	tap	0.25	20	7.7	4	99	1.1
(media grown)							
Acid-fast	BDF	0.3	25	7.0	60	40	>>60
Mycobacterium							
chelonei							
Virus							
Hepatitis A	BDF	0.5	5	10.0	49.6	99.99	12.3
Hepatitis A	BDF	0.5	5	6.0	6.5	99.99	1.8
Parasites							
G. lamblia	BDF	0.2–0.3	5	6.0	–	99	54–87
G. lamblia	BDF	0.2–0.3	5	7.0	–	99	83–133
G. lamblia	BDF	0.2–0.3	5	8.0	–	99	119–192

Note: [1)] C*t product of disinfectant concentration (C) in mg/l and [2)] BDF = buffered demand free [3)] ND = no datacontact time (t) in minutes for 99 per cent inactivation [4)] CDF = Chlorine demand free [5)] GAC = granular activated carbon

Potentially contaminated water such as coastal water or surface water should not be used at the production premises but may, if aesthetically acceptable, be used for removing waste material in places where no contact to food is possible.

A Water Quality Monitoring System

The responsible person should have continuously updated reference drawings of the pipe system and the authority to remove dead-ends. Especially in cases where a plant has undergone many changes, the piperuns may become more and more complicated over the years. The person should also be in contact with the local waterworks and the authorities in order to be informed of special events (repairs, pollution accidents or other changes).

A quality monitoring scheme could consist of a schematizised plan of all the sampling points and a checklist for each point describing what to examine and why, the frequency, who takes the sample, who does the analysis, what is the limit (value, tolerance) and what to do in case of deviation. If the water is obviously polluted there is of course no reason to wait for analytical results. The sampling frequency and the range of parameters will vary with the circumstances and the needs and possibilities of the specific plant. A minimum programme may for example consists of monitoring free chlorine daily and total counts plus coliforms on a weekly basis and a special, more intense monitoring programme to be used after repairs, when using new water supplies, etc.

The technical procedures describing the analyses for the common indicator organisms are given in standard textbooks. The WHO "Guidelines for drinking-water Quality", vol. 3 (WHO 1984b) mentions some methods and equipment suitable for small, rural supplies. The values used by the company should refer to the specific method employed and the recommendations should include how to sample (tap flow, volume, sampling vessel, labelling, etc.) and how to handle and examine the sample. Even though the commonly used methods for detecting, *e.g.*, fecal coliforms are standard analyses faulty handling of the samples often occurs. Samples should be processed within 24 hours or less and be kept cool, but not frozen (preferably below 5°C), and in the dark. The impact of sunlight can be very dramatic causing false negative results.

If chlorination is used for disinfection, monitoring of the free chlorine level is the simplest way of checking the water treatment and should be performed most often (*e.g.*, on a daily basis). Simple laboratory methods are described by WHO (1984b) and commercial dipsticks are now available for on-the-spot measurements (*e.g.* Merckoquant Chlor 100 from Merck). The microbiological indicator parameters may be checked less frequently. If disinfection systems leaving no residuals are being used, checking the equipment should be done regularly. The performance of the systems may be monitored at weekly intervals using indicator bacteria measurements.

CLEANING AND DISINFECTION

Cleaning and disinfection belong to the most important operations in today's food industries. Numerous and costly cases of food spoilage and unacceptable contamination with pathogenic bacteria has been traced back to failures or insufficiencies of these procedures.

The standards of hygiene required to avoid such problems are variable. In a plant, packaging products processed for safety (*e.g.*, by heat treatment) requirements will be very strict whereas handling of fresh chilled fish with a short shelf life and which is cooked before consumption, will be less demanding. Factors like housekeeping, personal hygiene, training and education, plant layout, design of equipment and machines, characteristics of materials selected, the maintenance and general condition of the plant can easily become more important than the actual cleaning and disinfection. For optimal use of resources and to ensure the microbiological quality of foods, it is important that all such factors are addressed when deciding on cleaning and disinfection procedures.

In some cases it may even be best to avoid cleaning and disinfection, because more harm than good can be done. As an example, this applies for dust accumulated on pipes and constructions unless time allows for a complete removal. Further, as another example, dry areas should always be kept dry and cleaning will then be limited to vacuuming if available, or sweeping, brushing, etc. It follows from the above that for each particular food plant or operation, implementation of cleaning and disinfection procedures is a project on its own where specialists, internal or external, should be consulted. Cleaning and disinfection will be processes like any other plant operation, and they should be equally documented and so should the corresponding process control, *i.e.*, the control of cleaning and disinfection respectively. If a HACCP concept is applied, these procedures should be treated as Critical Control Points (CCPs). If a Quality System like ISO 9000 is in operation, they should be integrated in the System. Responsible management realises that these procedures are integrated parts of production and poor hygienic condition in food processing plants will primarily be caused by management lack of knowledge and commitment.

For the whole process, three distinct operations are involved, *i.e.*

1. Preparatory work;
2. Cleaning and
3. Disinfection.

They are clearly distinct operations but linked firmly together in the way that the final result will not be acceptable, unless all three are carried out correctly. The various steps, which will be included in a complete cycle. Steps included in the complete cycle of preparatory work, cleaning, disinfection and control:

- Remove food products, clear the area for bins, containers, etc.
- Dismantle equipment to expose surfaces to be cleaned. Remove small equipment, parts and fittings to be cleaned in a specified area. Cover sensitive installations, to protect them against water, etc.
- Clear the area, machines and equipment for food residues by flushing with water (cold or hot) and by using brushes, brooms, etc.
- Apply the cleaning agent and use mechanical energy (*e.g.*, pressure and brushes) as required.
- Rinse thoroughly with water to completely remove the cleaning agent after the appropriate contact time, (residues may completely inhibit the effect of disinfection).
- Control of cleaning.
- Sterilisation by chemical disinfectants or heat.
- Rinse the sterilant off with water after the appropriate contact time. This final rinse is not needed for some sterilants, *e.g.*, H_2O_2, based formulations which decompose rapidly.
- After the final rinse, equipment is reassembled and allowed to dry.
- Control of cleaning and disinfection.
- In some cases it will be good practice to re-disinfect (*e.g.*, with hot water or low levels of chlorine) just before production starts.

Preparatory Work

The processing area is cleared of remaining products, spills, containers and other loose items. Machines, conveyors, etc., are dismantled so that all locations, where microorganisms can accumulate become accessible for cleaning and disinfection. Further electrical installations and other sensitive systems should be protected against water and the chemicals used.

Before use of the cleaning agent, a gross food debris removal procedure should be carried out by brushing, scraping or similar. All surfaces should be further prepared for the use of cleaning agents by a pre-rinse activity preferably with cold water which will not to coagulate proteins. Hot water may be used to remove fat or sugars in cases, where protein is not present in significant amounts. Completion of the preparatory work should be checked and recorded as any other process, to ensure the quality of the complete cycle of cleaning and disinfection.

Cleaning

Cleaning is undertaken to remove all undesirable materials (food residues, microorganisms, scales, grease etc.) from the surfaces of the plant and the process equipment, leaving surfaces clean, as determined by sight and touch and with no residues from cleaning agents. Microorganisms present will either be incorporated in the various materials or they attach to the surfaces as biofilms.

The latter will not be removed completely by cleaning, but experience has shown that a majority of the microorganisms will be removed. However, there will still be some left to be inactivated during the disinfection.

The effectiveness of a cleaning procedure in general depends upon:

- The type and amount of material to be removed.
- The chemical and physio-chemical properties of the cleaning agent (such as acid or alkali strength, surface activity, etc.) at the concentration, temperature and exposure time used.
- The mechanical energy applied, *e.g.*, turbulence of cleaning solutions in pipes, stirring effect, impact of water jet, "elbow-grease", etc.
- Condition of the surface to be cleaned.

Some surfaces, *e.g.*, corroded steel and aluminum surfaces can simply not be cleaned which means that disinfection also becomes very inefficient. The same applies for other surfaces, *e.g.*, wood, rubber, etc. The preferred material obviously will be high quality stainless steel.

The types of residues to be removed in food plants, will mainly be the following:

- Organic matter, such as protein, fat and carbohydrate. These are most effectively removed by strongly alkaline detergents (especially caustic soda, NaOH). Further, it is found that combinations of acid detergents (especially phosphoric acid) and non-ionic surfactants are effective against organic matter.
- Inorganic matter, such as salts of calcium and other metals. In beer

stone, milk stone, etc., salts are encrusted with protein residues. These are most effectively removed by acid cleaning agents.

- Biofilms, formed by bacteria, moulds, yeast and algae can be removed by cleaning agents that are effective against organic matter.

Most cleaning agents work faster and more effectively at higher temperatures, so it can be profitable to clean at a high temperature. Cleaning is often carried out at 60-80°C in areas where it pays energy wise to use such high temperatures.

Water

Water is used as a solvent for all cleaning and sterilising agents, and also for intermediate rinses and final rinse of equipment. The chemical and microbiological quality of the water is therefore of decisive importance for the efficiency of the cleaning procedures as already described in a previous section of this chapter. In principle, water used for cleaning must be potable. Hard water contains a large amount of calcium and magnesium ions. When the water is heated, calcium and magnesium salts corresponding to the temporary hardness will precipitate as insoluble salts. Also, some cleaning agents, especially alkalis, can precipitate calcium and magnesium salts. Apart from reducing the effectiveness of detergents hard water leads to the formation of deposits or scales.

Scales which can be formed in several other ways are not only unsightly but objectionable of several reasons:

- They harbour and protect microorganisms.
- They reduce the rate of heat exchange on heat exchanger surfaces. This could lead to underprocessing, underpasteurisation or understerilisation.
- The presence of scales tends to increase corrosion.

The formation of scales can be reduced by addition of chelating and sequestering agents, which bind calcium and magnesium in insoluble complexes. However, it is advisable to prevent precipitations by softening the water before it is used for cleaning. Softening can be effectively achieved by ion exchange, in which the calcium and magnesium ions are replaced by sodium ions, the salts of which are soluble. A modern, and more costly, method of softening water is by means of reverse osmosis. Microbiological purity of water to be used for final rinse must be beyond reproach. If not, it will in some cases be acceptable to include low levels of chlorine, *i.e.*, a few ppm.

Cleaning Agents

The ideal detergent would be characterised by the following properties:

- It possesses sufficient chemical power to dissolve the material to be removed.

- It has a surface tension low enough to penetrate into cracks and crevices; it should be able to disperse the loosened debris and hold it in suspension.
- If used with hard water, it should possess water softening and calcium salt dissolving properties to prevent precipitation and build-up of scale on surfaces.
- It rinses freely from the plant, leaving this clean and free from residues, which could harm the products and affect sterilisation negatively.
- It does not cause corrosion or other deterioration of the plant. It is recommended always to check by consulting the supplier of machines, etc.
- It is not hazardous for the operator.
- It is compatible with the cleaning procedure being used, whether manual or mechanical.
- If solid, it should be easily soluble in water and its concentration easily checked.
- It complies with legal requirements concerning safety and health as well as biodegradability.
- It is reasonably economical to use.

A detergent with all these characteristics does not exist. So one must, for each individual cleaning operation, select a compromise by choosing a useable cleaning agent and water treatment additives so that the combined detergent has the properties that are most important for the procedure concerned.

When choosing a cleaning agent, one can pick either a ready-mixed factory product, which has the desired properties, or it can be homemade. In this case it must be assured that the components are mutually compatible.

Cleaning Systems

The various steps and including sterilisation, represents the most comprehensive procedure for manual cleaning and disinfection or Clean Out of Place (COP). It is suitable for modern plants. For cleaning liquid handling plants like breweries and dairies Clean In Place (CIP) Systems will be used, based on circulation by pumping of water, cleaning agents and disinfectants. In principle the two systems will be similar.In most factories, a combination of COP and CIP will be used. Use of CIP may be limited to part of the plants or even to a particular machine. However, regardless of the type and size of food production the general principles behind the complex cycle should be kept in mind and applied to ensure effective cleaning and disinfection.

The frequency of cleaning and disinfection will vary from several times during the working day, *i.e.*, at every major break to once every day, at end of production, or even less frequent. Sometimes disinfection will not be included,

Table. Shows Important Characteristics of the Cleaning Agents Most Commonly used in the Food Industry.

Categories of aqueous cleaners	Approximate concentrations for use (%, w/v)[1)]	Examples of chemical used[2)]	Functions	Limitations
Clean water	100	Usually contains dissolved air and soluble minerals in small amounts	Solvent and carrier for soils, as well as chemical cleaners	Hard water leaves deposit on surfaces. Residual moisture may allow microbial growth on washed surfaces.
Strong alkali	1–5	Sodium hydroxide Sodium orthosilicate Sodium sesquisilicate	Detergents for fat and protein. Precipitate water hardness	Highly corrosive. Difficult to remove by rinsing. Irritating to skin and mucous membranes.
Mild alkali	1–10	Sodium carbonate Sodium sesquisilicate Trisodium phosphate Sodium tetraborate	Detergents. Buffers at pH 8.4 or above Water softeners	Mildly corrosive. High concentrations are irritating to skin
Inorganic acid	0.5	Hydrochloric Sulphuric Nitric Phosphoric Sulphamic	Produce pH 2.5 or below Remove inorganic precipitates from surfaces	Very corrosive to metals, but can he partially inhibited by anti-corrosive agents. Irritating to skin and mucous membranes
Oragnic acids	0.1–2	Acetic Hydroxyacetic Lactic Gluconic Citric Tartaric Levulinic Saccharic		Moderately corrosive, but can be inhibited by various anti-corrosive compounds
Anionic wetting agents	0.15 or less	Soaps Sulphated alcohols Sulphated hydrocarbons Aryl-alkyl polyether sulphates Sulphonated amides Alkyl-arylsuphonated	Wet surfaces Penetrate crevices and woven fabrics Effective detergents Emulsifiers for oils, fats, waxes, and pigments Compatible with acid or alkaline cleaners and may be synergistic	Some foam excessively Not compatible with cationic wetting agents
Non-ionic wetting agents	0.15 or less	Polyethenoxyethers condensates Amine-fatty acid condensate	Excellent detergents for oil. Ethylene oxide-fatty acid agents to control foam	May be sensitive to acids Used in mixtures of wetting
Cationic wetting agents	0.15 or less	Quaternary ammonium	Some wetting effect Antibacterial action	Not compatible with anionic wetting agents
Sequestering agents	Variable (depending on hardness of water)	Tetrasodium pyrophosphate Sodium tripolyphosphate Sodium hexametaphosphate Sodium tetrapolyphosphate Sodium acid pyrophosphate Ethylenediaminetetra-acetic acid (sodium salt) Sodium gluconate with or without 3% sodium hydroxide	Form soluble complexes with metal ions such as calcium, magnesium and iron to prevent film formation on equipment and utensils See also strong and mild alkalis above	Phosphates are inactivated by protracted exposure to heat Phosphates are unstable in acid Solution

Abrasives	Variable	Volcanic ash Seismotite Pumice Feldspar Silica flour Steel wool[3] Metal of plastic 'chlore balls[3] Scrub brushes	Removal of dirt from surfaces with scrubbing Can be used with detergents for difficult cleaning jobs	Scratch surfaces Particles may become imbedded in equipment and later appear in food Damage skin of workers
Chlorinated compounds	1	Dichlorocyanuric acid Trichlorocyanuric acid Dichlorohydantoin	Used with alkaline cleaners to petizing of proteins and minimize milk deposits	Not germicidal because of high pH Concentrations vary depending on the alkaline cleaner and conditions of use
Amphoterics	1.2	Mixtures of a cationic amine salt or a quaternary ammonium compound with an anionic carboxy compound, a sulfate ester, or a sulfonic acid	Loosen and soften charred food residues on ovens or other metal and ceramic surfaces	Not suitable for use on food contact surfaces[4]
Enzymes	0.3 -1	Proteolytic enzymes	Digest proteins and other complex organic soils	Inactivated by heat Some people become hyper-sensitive to the commercial preparations.

Note: 1) Concentration of cleaning agent in solution as applied to equipment

2) Some regulatory agencies require prior approval

3) Steel wool and metal 'chlore balls' should not be used on food plant

4) Some amphoteric disinfectants are used on food contact surfaces

e.g., in areas to be kept dry and for environments with materials which cannot be disinfected or premises unsuitable for disinfection. In such cases cleaning is still very important for the general appearance and hygienic condition of the plant or premises and the general attitude towards hygiene of the employees.

Control of Cleaning

The effective cleaning is a prerequisite for an efficient disinfection. This indicates the importance of controlling cleaning. The most important control is visual inspection and other rapid tests to demonstrate the following important results of cleaning:

- That all cleaned surfaces are visibly clean.
- That all surfaces by feeling are free from food residues, scales and other materials and by smelling free from undesirable odours.

Further, the concentrations and pH-values of cleaning agents, temperatures, if hot cleaning is used, and contact times should be monitored and registered. pH measurements, or similar testing, of rinse water may be used to ensure that the cleaning agent is removed so that it will not interfere with the disinfectant.

These controls are all rapid and allow immediate decisions to be made as to whether cleaning should be repeated, partly or completely, or to proceed to the process of disinfection. All controls, etc., shall be registered as part of the Quality System. At this stage, microbiological control serves no real purpose. Firstly biofilms and surviving microorganisms are likely to be present and secondly, reliable rapid methods are not available.

Disinfection

Traditionally, the terms "disinfection" and "disinfectants" are used to describe procedures and agents used in food industries to ensure a microbiologically acceptable standard of hygiene. This practice will be followed although it is realised that the procedures and agents described will rarely introduce 'sterility' *i.e.*, total absence of viable microorganisms. Disinfection can be effected by physical treatments such as heat, U.V. irradiation, or by means of chemical compounds. Among the physical treatments, only heat shall be described. The use of heat in the form of steam or hot water is a very safe method and a widely used method of disinfection. The most commonly used chemicals for disinfection are:

- Chlorine and chlorine compounds.
- Iodophors.
- Peracetic acid and hydrogen peroxide.
- Quaternary ammonium compounds.
- Ampholytic compounds.

Disinfection by Use of Heat

Heating at suitably high temperatures for a suitably long time is the safest method for killing microorganisms. The velocity with which heat killing occurs depends on temperature, humidity, type of microorganism and the environment in which the microorganisms occur during heat treatment. If microorganisms are entrapped in scales or other substances, they are protected and not even heating may be effective. It is important to recall the kinetics for heat inactivation of microorganisms:

$$\log C_t = \log C_o - K \times t,$$

where C_o = original population of living microorganisms (initial viable count) and C_t = total surviving after time t. K is a constant (= slope of the straight line) and depends on the microorganism concerned and the experimental conditions. K is described as the death rate. It is seen that the number of surviving microorganisms at time 't' is determined by the initial level of infection, as well as the death rate constant and the heating time. Circulation of hot water (about 90°C) is very effective. The water should be circulated for at least 20 minutes after the temperature of the return water has risen to 85°C or more. Obviously steaming is equally effective when applicable.

Disinfection by Use of Chemical Agents

With the use of chemical disinfectants, the death rate for microorganisms depends, among other things, upon the agent's microbicidal properties, concentration, temperature and pH as well as the degree of contact between disinfectant and microorganisms. Good contact is obtained, *e.g.*, by stirring, turbulence, smooth surfaces and low surface tension.

Table. Comparison of the More Commonly used Disinfectants (ICMSF 1988)

		Steam	Chlorine	Iodophores	QAC/QUATS surfactants	Acid anionic
Effective against	Gram-positive bacteria (lactics, clostridia, *Bacillus, Staphylococcus)*	Best	Good	Good	Good	Good
	Gram-negative bacteria (*E. coli, Salmonella, psychrotrophs)*	Best	Good	Good	Poor	Good
	Spores	Good	Good	Poor		Fair
	Bacteriophages	Best	Good	Good		Poor
Properties	Corrosive	No	Yes	Slightly	No	Slightly
	Affected by hard water	No	(No)	Slightly	Some are	Slightly
	Irritative to skin	Yes	Yes	Yes	No	Yes
	Affected by organic matter	No	Most	Somewhat	Least	Somewhat
	Incompatible with:	Materials sensitive to high temperature	Phenols, amines, soft metals	Starch, silver	Anionic wetting agents, soaps.	Cationic surfactants and alkaline detergents
	Stability of use solution		Dissipates rapidly	Dissipates slowly	Stable	Stable
	Stability in hot solution (greater than 66°C)		Unstable, some compounds stable	Highly usable (best used below 45° C)	Stable	Stable
	Leaves active residue	No	No	Yes	Yes	Yes
	Tests for active residue chemical	Unnecessary	Simple	Simple	Simple	Difficult
	Maximum level permitted by USDA and FDA w/o rinse	No limit	200 ppm	25 ppm	25 ppm	
	Effective at neutral pH	Yes	Yes	No	No	No

As with heat disinfection, different microorganisms show different resistance to chemical sterilants. It is also so that contamination by inorganic or organic matter can reduce the death rate considerably. As mentioned before an effective disinfection can only be obtained after an effective cleaning.

The desirable plant disinfectant would be characterised by the following properties:

- It has sufficient anti-microbial effect to kill the microorganisms present in the available time and should have a sufficiently low surface tension to ensure good penetration into pores and cracks.
- It rinses freely from the plant, leaving this clean and free from residues which could harm the products.
- It must not lead to development of resistant strains or any surviving microorganisms.
- It does not cause corrosion or other deterioration of the plant. It is recommended that the suppliers of machines, etc., be asked before chlorine or other aggressive disinfectants are taken into use.

- It is not hazardous to the operator.
- It is compatible with the disinfection procedure being used, whether manual or mechanical.
- If solid, it should be easily soluble in water.
- Its concentration is easily checked.
- It is stable for extended storage periods.
- It complies with legal requirements concerning safety and health as well as biodegradability.
- It is reasonably economical in use.

It will often be necessary to combine sterilants with additives in order to obtain the required properties. To prevent development of resistant strains of microorganisms it can be advantageous to change from one type of sterilant to another from time to time. This is especially advisable when quaternary ammonium compounds are used. Among the most used sterilants the following shall be described briefly.

Chlorine is one of the most effective and widely used disinfectants. It is available in several forms like sodium hypochlorite solutions, chloramines and other chlorine containing organic compounds. Gaseous chlorine and chlorine dioxide are also used. Chlorinated sterilants at a concentration of 200 ppm free chlorine are very active and also with some cleaning effect. The disinfectant effect is considerably decreased when organic residues are present.

The compounds dissolved in water will produce hypochlorous acid, HOCl, which is the active sterilising agent, acting by oxidation. In solution it is very unstable, particularly in acid solution where oxic chlorine gas will be liberated. Furthermore, solutions are more corrosive at low pH. Unfortunately, the germicidal activity is considerably better in acid solution than in alkaline, thus the working pH should be chosen as a compromise between efficiency and stability. Organic chlorinated sterilants are generally more stable but require longer contact times.

When used in the proper range of values (200 ppm free chlorine), chlorinated sterilants in solutions at ambient temperatures are non-corrosive to high quality stainless steel but they are corrosive to other less resistent materials. Iodophors contain iodine, bound to a carrier, usually a non-ionic compound, from which the iodine is released for sterilisation. Normally the pH is brought down to 2-4 by means of phosphoric acid. Iodine has its maximum effect at this pH range. Iodophors are active disinfectants with broad antimicrobial spectrum just like chlorine. They are inactivated by organic material. Concentrations corresponding to approx. 25 ppm free iodine will be effective.

Commercial formulations are often acidic making them able to dissolve scales. They can be corrosive depending on the formulation and they should not be used above 45°C as free iodine may be liberated. If residues of product and caustic cleaning agents are left in dead legs and similar places, this may in combination with iodophors cause very unpleasant "phenolic" off-flavours. Hydrogen peroxide and peracetic acid are effective sterilants acting by oxidation and with a broad antimicrobial spectrum. Diluted solutions may be used alone or in combination for disinfection of clean surfaces. They lose their activity more readily than other sterilants in the presence of organic substances and they rapidly loose their activity with time.

Quaternary ammonia compounds are cationic surfactants. They are effective fungicides and bactericides but often less effective against Gram negative bacteria. To avoid development of resistant strains of microorganisms, these compounds should only be used alternating with the use of other types of disinfectants. Due to their low surface tension, they have good penetrating properties and for the same reason, they can be difficult to rinse off. If quats come into contact with anion-active detergents, they will precipitate and become inactivated. Mixing or successive use of these two types of chemicals must therefore be avoided. Ampholytic sterilants have properties similar to quaternary ammonia compounds.

Control of Disinfection

Control of disinfection will be the final control of the complete cycle of cleaning and disinfection.

Provided cleaning has been controlled effectively as described above control of disinfection will be effective when the following conditions are met:

- Control of time and temperature conditions for disinfection by heat.
- Control of active concentrations of chemical disinfectants.
- Control that all surfaces to be disinfected are covered by the disinfectant.
- Control of contact time.

The above controls should be documented and the observations reported and registered as required in standard Quality Systems. Microbiological testing and control serve the purpose of verification. Various techniques are available but none are ideal and they are not "real time" methods which is highly desirable for control of cleaning and disinfection. Overnight incubation is too late to correct critical situations. However, if conducted at regular intervals and planned to cover all critical points, useful information from microbiological control can be accumulated with time.

Various methods are used and shall be mentioned briefly.

- *Swab testing*: This is the most usual technique and one of the better ones. By use of a sterile swab of cottonwool, part of the disinfected surface is swabbed, and the bacteria transferred to the swab is transferred to a diluent for determination of colony forming units in standard agar substrates. Swabs are especially useful in places, where other control methods can only be used with difficulty, *i.e.*, pockets, valves, etc.

- *Final rinse water*: Membrane filtration of rinse water and incubation on agar substrate is a very sensitive technique for control of CIP systems as well as other cleaning and disinfection systems, where a rinse can be applied.
- *Direct surface plates*: In these methods petri dishes or contact slides with selective or general purpose agar media are applied to the surface to be examined, followed by incubation and counting of colony forming units. These techniques can only be applied to plane surfaces, which is a limiting factor.
- *Bioluminometric assay of ATP*: This is almost a "real time" method giving the answer within minutes. It is very sensitive and can be combined with swabbing for collection of microorganisms from surfaces. The method is rather non-specific, and it may not be able to distinguish between microorganisms and food residues. However, if applied under defined conditions it may prove useful and superior to the conventional methods, because it provides the answer in minutes.

Regardless of the technique used, it is valuable to know from the verification analyses that the system was working, when it was established. There is also a value in knowing trends as expressed in the verification results recorded. The objective of studying trends and conducting the microbiological control of cleaning and disinfection, obviously will be, to take corrective action before loss of control of products or processes occur.

HANDLING OF FRESHWATER FISH BEFORE PROCESSING

The quality of the raw material and its usefulness for further utilisation in processing is affected by the fish capture method. Unsuitable fishing methods, *e.g.*, catching too many fish in one haul, cause not only mechanical damage to the fish, but also create stress and the conditions which accelerate processes which begin after fish death.

In many countries consumers are used to buying live fish: this assures the highest quality. This habit takes different forms, *e.g.*, the consumer buys live fish, for instance carp or trout and processes it at home. Very often the fish bought live can be partly processed by the shop assistant; for example, it can be filleted. In some restaurants the customer can choose the fish from an aquarium and have it prepared for consumption. Thus the tradition, the quality, and the resultant price, constitute the reason why the preparation of fish for transportation, and the transportation itself, are the preliminary operations of processing of freshwater fish like trout, carp, eel, etc. However, producers should remember that not all fish are suitable for transportation alive. Therefore, just after fishing, fish should be sorted and only those in good condition, healthy and not damaged be destined for sale as live fish. Fish so classified is first conditioned in water of appropriate quality. The conditioning process reduces stress, inhibits metabolism and at the same time food remains are removed from the alimentary ducts and the oxygen demand reduced.

During the conditioning process fish is not fed which further inhibits metabolism and also limits the excretion of ammonia and carbon dioxide. In the short conditioning process 1 m^3 of water is sufficient for 50-60 kg of carp, 30-40 kg of pike, 20-25 kg of trout or pike-perch. Water provided for conditioning must be properly oxidised. For example, in the case of 1 kg of fish at a temperature of 10° C the oxygen demand is: eel 25 mg, carp 45 mg, pike 50 mg. Young fish need more oxygen than older fish. Oxygen consumption depends also on the liveliness of fish. The amount of oxygen dissolved in water depends on water temperature which should be rather low. But for stenothermal species such as carp water temperature should be not less than 10-12° C in summer and 5-6° C in spring and autumn. Optimal temperature for conditioning and transportation of trout is 5-6° C in summer and 3-5° C in spring. During winter fish tolerates temperatures of 1-2° C.

Nowadays, special tanks with aeration system and often with cooling and filtering (activated coal, biological filters) systems are used for transportation of live fish. In simple solutions water is cooled by ice. Cooling is especially important during summer and in transportation over long distances. If all parameters, *i.e.*, temperature, oxygenation, are properly maintained, and when the temperature does not exceed 10° C, the weight loss varies from 1 to 6 per cent, and about 10 per cent of carp and 20 per cent of trout die during a six-day transportation in winter. At present, large valuable fish species are transported via air in which case they are placed in big plastic bags with aeration system.

Equipment for Preliminary Processing of Freshwater Fish

Preliminary processing of freshwater fish usually consists of the following steps or unit processes: evisceration, deheading, scaling, cutting of fins and belly flaps, slicing of whole fish into steaks, filleting, skinning, grinding of skinned fillets and different combinations of the above.

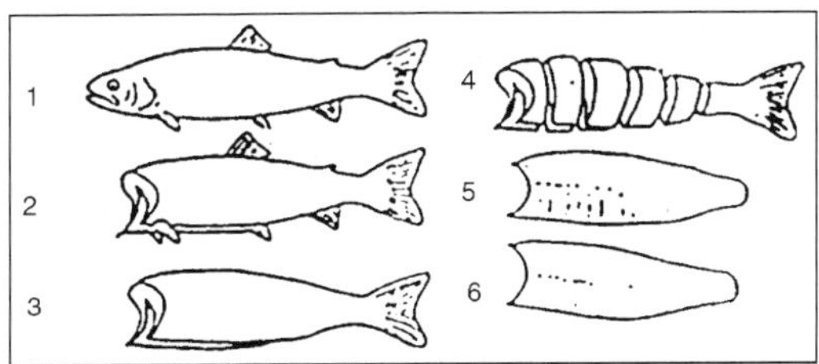

Fig. Major Forms of Preprocessed Fish: 1. Whole fish, 2. Gutted fish without head, 3. Gutted fish without head and fins, 4. Sliced whole fish after deheading and evisceration, 5. Fillet with ribs, and 6. Fillet without ribs, with or without the skin.

The products of preliminary processing can be sold or further processed to obtain value added products. In freshwater fish processing, particularly species such as perch, pike-perch and the cyprinids, the processing steps described above are executed manually with a wide variety of knives.

Efficient preparation of fish is important when top quality, maximum yield and highest possible profits are to be achieved. This is important when fish is to be exported. Efficient fish preparation is a skill only be acquired with practice.

Several perfectly acceptable methods for cutting any fish exist; they may often give the same yield and similar end-products. In the future, the level of mechanisation of fish processing in small processing plants will increase due to the constant pressure to reduce production costs and improve economic performance.

The present level of mechanisation is low which results from the overall limited production, seasonal availability of the raw product and lack of inexpensive, efficient mechanical equipment adaptable for processing of various fish species.

In practice, most freshwater fish processing is done in small processing plants (with the exception of salmon and trout processing), usually supplying products for local or nearby markets. Manpower capacity in such plants varies, usually not exceeding 10-20 employees. In addition to freshwater fish, frozen marine fish may be processed in the same plant.

Stunning of Fish

In many freshwater species the method of stunning is critical for final product quality because prolonged agony of fish causes production of undesired substances in the tissue. Oxygen deficiency in blood and muscle tissue results in accumulation of lactic acid and other reduced products of catabolic processes and consequently in a paralysis of the neural system. Red spots appear on the surface of the skin and in the muscle tissue near the backbone; these reduce quality.

Stunning of freshly caught fish or fish delivered live to a processing plant is best done with an electric current. First, the fish are placed in a tank of water and an electric current is then passed through the water to stun or kill the fish. Live fish are also slaughtered by cutting the aorta and bleeding to death when technological or ritual reasons require the removal of blood from the tissue before further processing. In some plants, water in the fish tanks is saturated with carbon dioxide which renders the animals unconscious or dead.

Grading Processing

The processing sequence starts from grading the fish by species and size. Sorting by species or on the basis of freshness and physical damage are still manual processes, but grading of fish by size is easily done with mechanical equipment. Mechanical graders yield better sorting precision for fish before or after rigour mortis than for fish in a state of rigour mortis. Size grading is very important for fish processing (*i.e.*, smoking, freezing, heat treatment, salting, etc.) as well as for marketing. Automated sorters are rarely used in small plants processing freshwater fish because the raw product is usually already sorted on delivery and because of their high costs.

Automated grading is 6-10 times more efficient than manual grading. The sorting speed of different graders varies and depends on the type of device and size of fish sorted. Sorting capacity is 1-15 t/hour, and usually into three size groups.

A combination of conveyor belt and automated sorter is used by fish processing plants in the USA. This machine has an interesting design: two smooth rotating rollers are installed above the surface of the conveyor belt and the distance between the rollers and belt can be adjusted according to the maximum thickness of the sorted fish.

Thinner animals fall off the belt while the thick ones are retained on it until the end of line. Therefore, one device serves simultaneously as a grading machine and a conveyor.

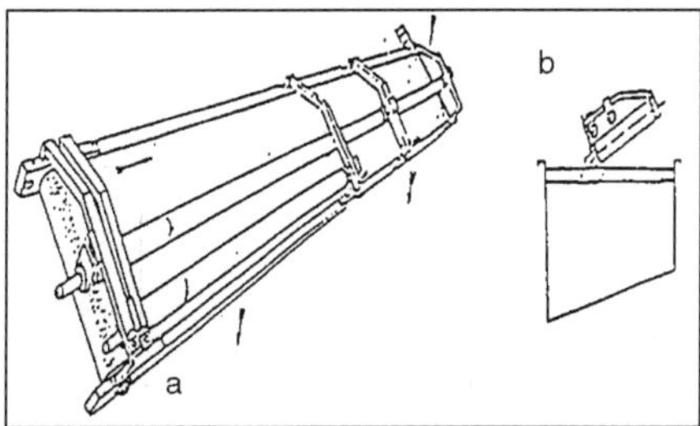

Fig. Combination Grading Machine-conveyor Belt: a - General View, b - Cross-section

Most commonly used grading machines consist of a series of compartments connected by slits of varying size with rotating rollers or conveyor belts arranged in a V-shape. In such devices fish are sorted according to the maximum thickness which is highly correlated to fish length. The size range to be sorted is easily adjusted.

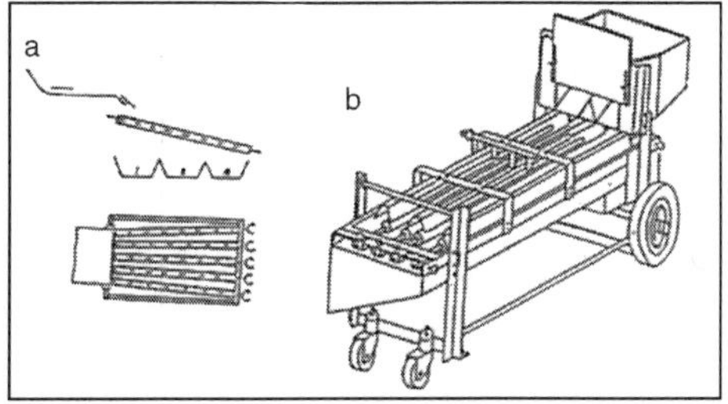

Fig. Grading Machine with a Fan Shaped Arrangement of Rollers: a - Scheme, b - General View.

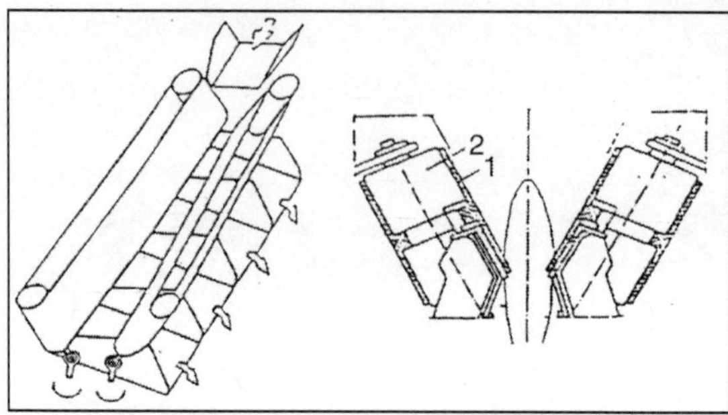

Fig. Slit Grader Consisting of Two Conveyor Belts Arranged in a V-shape; 1 - Rubber Belt, 2 - Rotating Wheel

Removal of Slime

Slime accumulating on the skin surface of dying fish is a protection mechanism against harmful conditions. In some freshwater species slime constitutes 2-3 per cent of body weight. Slime excretion stops before rigour mortis. Slime creates a perfect environment for micro-organism growth and should be removed by thorough washing.

Eel, trout and carp require special care with regard to slime removal. Even small amounts of slime, which frequently remain after manual cleaning, result in visible yellowish-brown spots (particularly in smoked eel). Drum-washing with a horizontal rotation axis does not remove slime from some fish, *e.g.*, eel. Eel are best washed in machines which originally serve as scalers. The device is loaded with 30 kg of eel and several kilograms of salt, and after about 2-3 minutes the slime is completely removed from the fish skin. This procedure is more efficient than manual washing.

Slime can be removed from eel, trout and other freshwater species by soaking fish in a 2 per cent solution of baking soda and then washing in a cylindrical rotating washer.

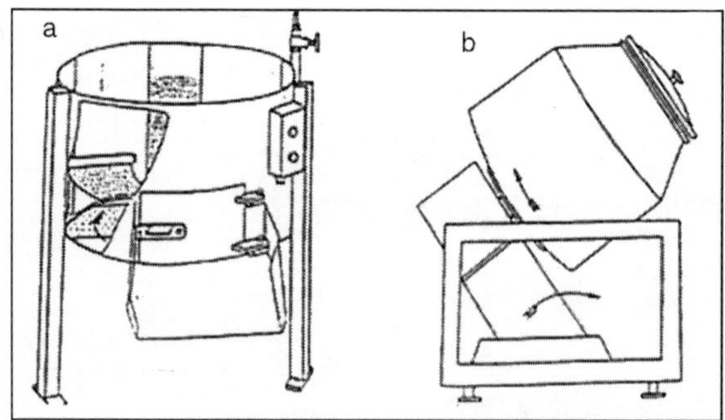

Fig. a. Vertical Drum Scaler with Rotating Bottom, b. Rotating Cylindrical Scaler with a Tilted Axis of Rotation (Adapted form a Cement Mixer). Both Machines can be Used for Slime Removal.

Scaling

Many freshwater species are routinely scaled; this is extremely labour-intensive when done manually. Some sources estimate that manual scaling of larger animals requires almost 50 per cent of the total time necessary to produce headed and gutted fish without fins. Fish destined for skinning and filleting or to be smoked or minced in mincing/deboning separator is not scaled. Tools used for manual scaling are shown in Figure: Tools are moved over the body of fish from tail fin towards the head, pulling out the scales.

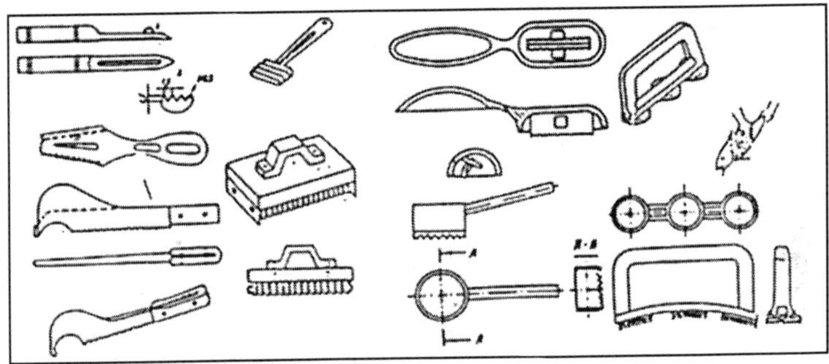

Fig. Tools Used for Manual Scaling.

Fish such as perch, bream, pike-perch and carp, are particularly difficult to scale manually. One method includes blanching of fish for 3-6 seconds in boiling water and then scaling by hand with motions perpendicular to the long body axis. Mechanised and power-assisted hand-held scalers are commonly used in small processing plants.

Electrical hand-held scalers simplify and speed up the scaling procedure. They are most commonly used for secondary scaling of fish which has left the automated scaling device 80-90 per cent free of scales.

Use of electrical hand-held scalers reduces labour intensity and assures complete elimination of scales. The power-assisted tool consists of a cylindrical rotating scraper of 30-40 mm diameter powered by an electric motor and connected to it with a flexible rod. The vertical cylindrical scaler with rotating bottom and fixed side wall is widely used in small fish processing plants.

Fish (usually 30-40 kg) is loaded from the top and unloaded through the door in the side wall. Scales catch on small contoured slits cut in the bottom and side wall of the device, and are thus pulled out of the skin. The same machines can be used for slime removal.

Cement mixers are often utilised for scaling after the original cylinder is replaced with a 120-l drum made of stainless steel, with punctured contoured slits of 10 mm diameter.

In addition to devices which have been specifically designed for scaling, a variety of automated tools can be employed, *e.g.*, vegetable peelers. However, their use may result in mechanical damage to the fish even after modifications.

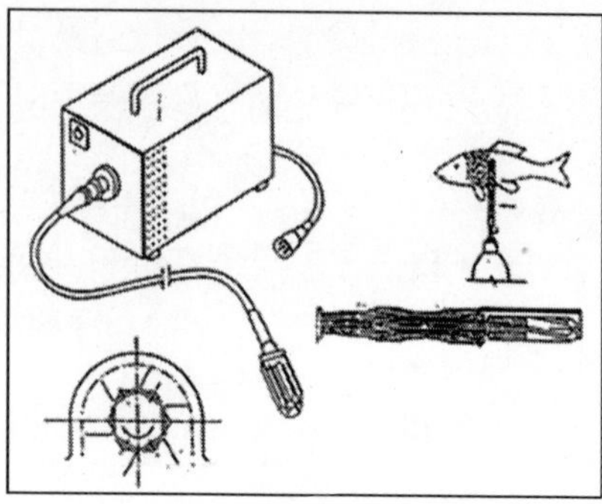

Fig. Electrical Scaler.

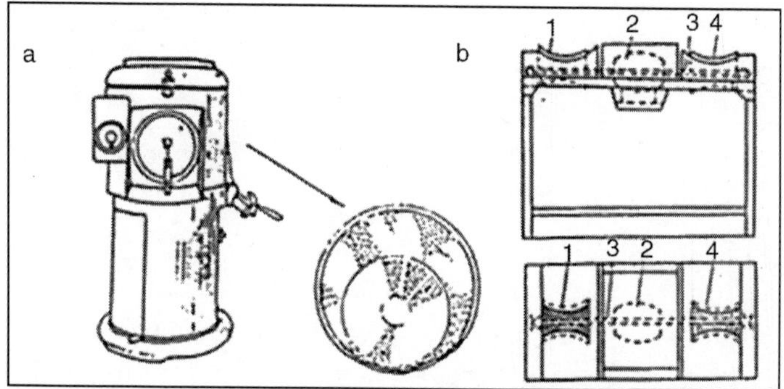

Fig. a. Scaler Modified from a Mechanical Vegetable Peeler (Rotating Bottom), b. Drum Scaler with Two Processing Lines; 1,4 - Rough Surface Rotating Drum, 2- Engine, 3- Drive Shaft.

A semi-automated device is used for scaling larger fish; fish is manually passed over the rough surface rotating drums which have contoured slits of 3-4 mm depth. One worker can scale 10-20 fishes/minute (scaling speed varying with species). Special protective gloves must be worn during this procedure.

Various scalers are designed on the same principle. The processing time of a cylindrical rotating scaler with the horizontal rotation axis is from 2 to 7 minutes depending on the species and size as well as on the type of slits on the surface of the drum and the rotational speed. The total weight of fish loaded in one run rarely exceeds 30-60 kg.

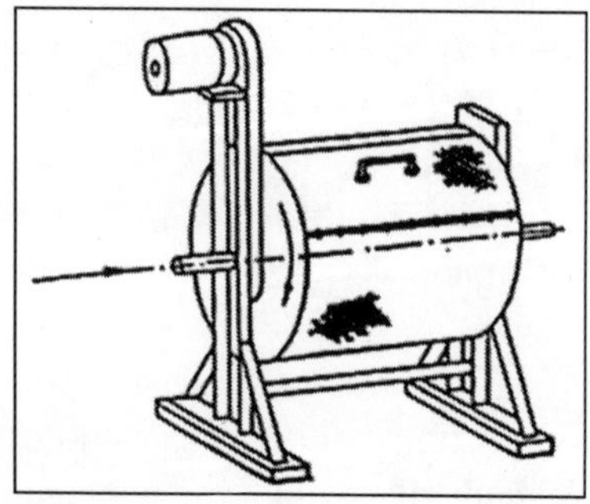

Fig. Cylindrical Scaler with Horizontal Rotation Axis.

Another kind of cylindrical scaler with a horizontal rotation axis can be periodically tilted during a scaling cycle which causes fish to tumble inside the drum, and consequently scales more efficiently. In some fish species, the scales can be removed from fish with a pressurised stream of water while fish is placed inside the scaler drum. The drums of such devices are made either of stainless mesh with rough edges or of stainless sheets perforated with contoured slits which detach the scales. Water has to be injected into the drum for the machine to operate. Less common are cylindrical scalers with a continuous operating cycle.

Washing

Washing is intended primarily to clean the fish and to remove accumulated bacteria. The effectiveness of the washing procedure depends, *inter alia*, on the kinetic energy of the water stream, ratio of fish volume to water volume and on the water quality. A proper fish:water volume ratio for achieving the desired level of cleanliness is 1:1, however, in practice more water is usually used (twofold).

Washing of gutted and headed fish should be done on termination of the processing operation. To improve the effectiveness of the cleaning procedure, various mechanised scrubbing devices are utilised which can remove up to 90 per cent of the initial bacterial contamination. Potable water is used for washing in freshwater fish processing plants.

The following washers are commonly used: vertical drum, horizontal drum and a combination washer-conveyor belt. The operation cycle for these machines is 1-2 minutes. The vertical drum washer is frequently used because of its conveniently small size. The most common is the horizontal tumbler washer. A rotating perforated drum constitutes the main component of this device; the drum is usually 2-4 m long, with round holes 10 mm in diameter. Inside the drum there are metal or rubber bars which facilitate tumbling and mixing of fish. Rotation of the drum, its tilted axis and the arrangement of internal bars result in a movement of fish towards the outlet of the device. Washing is continuous and is accomplished by spraying pressurised water through the perforated pipe installed inside the drum. Dirty water collects in the waste basins.

The mechanised washers described can be used to process whole fish, deheaded and gutted fish as well as boneless fillets because the washing action generates no physical damage to the product. Due to their continuous operating cycle, horizontal-axis drum washers are particularly suitable for production lines requiring constant product flow. A combination washer-conveyor is less popular but can serve to separate fish from ice: ice, having lower density than water, floats to the water surface from where it is removed, while fish falls onto the meshed conveyor and leaves the washing basin. Although there is an additional water jet at the exit from the water basin, the effectiveness of washing in this washer is lower than in the drum washers; fish on the conveyor belt is not exposed to scrubbing which is so important in the tumbler washers. The meshed conveyor (stainless steel or plastic mesh) with a water spraying system.

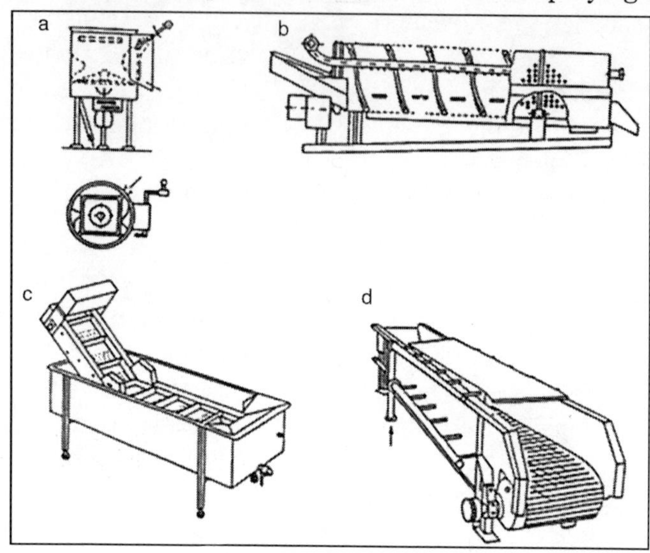

Fig. a. Vertical-axis Drum Washer, b. Horizontal-axis Drum Washer, c. Combination Washer-conveyor Belt, d. Conveyer with a Water Spraying System (Belt Made of Metal Mesh) Used as a Washer.

Deheading

The head constitutes 10-20 per cent of the total fish weight and it is cut off as an inedible part. Although many mechanised deheading machines had been developed for processing marine fish, freshwater fish are usually deheaded manually. The main reason is the lack of inexpensive equipment offering minimal tissue loss during this procedure. Different cutting techniques used for deheading.

A cut around the operculum, a so-called round cut, results in lowest meat loss. This technique is 4-5 per cent more efficient than the straight cut commonly used in mechanised systems. The contoured cut, which runs perpendicular to the fish's backbone and then at an angle of 45o, is also advantageous. This particular deheading technique is used when fillet, mainly boneless and skinned, is the final product. The head is removed with the pectoral bones and fins.

In small freshwater fish processing plants, small fish are frequently deheaded manually. Deheading of larger fish requires much more effort and automated heading devices are essential. Unfortunately, a single deheading machine which would cover a broad spectrum of fish sizes, *i.e.*, 20-110 cm, does not exist. An average deheading device can usually be used to process fish for which a difference between minimum and maximum length does not exceed 30-40 cm.

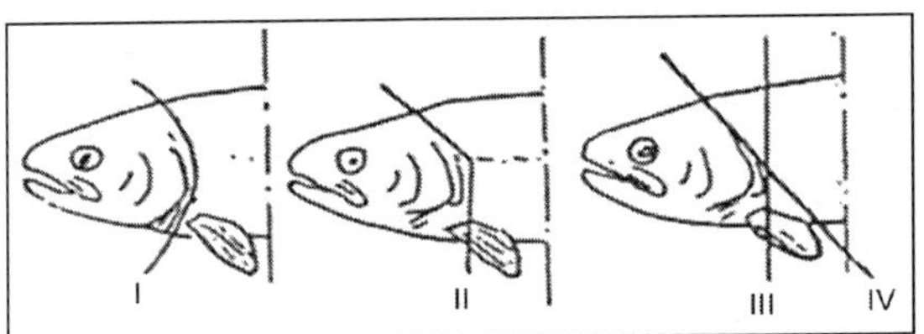

Fig. Cutting Techniques Used for Deheading of Freshwater Fish: I - Round Cut, II - Contoured Cut, III - Straight Cut, IV - Slant Cut.

The cutting elements used in the deheading machines are either disc, contoured, cylindrical knives, band saws or guillotine cutters. A machine operator adjusts the position of the cutting element according to the fish size. Thus the amount of meat lost during the deheading procedure depends not only on the type of head cut but on the experience and skill of the operator. The speed of a deheading device depends on the size of fish processed and is usually 20-40 fish/minute. In some plants, simple - and sometimes rigged by an amateur - deheading devices are used which can potentially cause severe physical damage to the operator's hands.

It is very important to examine safety problems associated with handling of the device before making a final decision about its purchase. The deheading machine with a guillotine cutter is used for deheading larger freshwater fish; cutters are changed according to species and size range. Economical cuts such as contoured cut or cut around operculum can be performed by changing the cutters. In one type of deheading device with cylindrical rotational saw the round cut is used. The most commonly utilised saw sizes are 12, 15 and 18 cm in diameter; saw size is adjusted to the fish species and size. The simplest designs are represented by the deheading machines with a circular saw and with a disc saw which also acts as a guillotine.

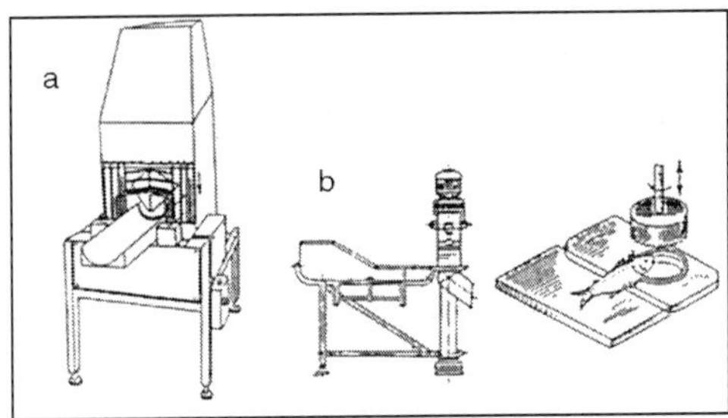

Fig. a. Deheading Machine with Guillotine Cutter, Suitable for Larger Fish (Contoured or Round Cut), b. Deheading Machine with a Cylindrical Rotational Saw (Round Cut).

Gutting

The purpose of gutting is to remove those fish body parts most likely to reduce product quality, as well as to remove gonads and sometimes the swim bladder. Evisceration of freshwater fish is labour-intensive and usually performed by hand. Gutting consists of cutting down the belly (fish may be deheaded or not), removal of internal organs, and, optionally, cleaning the body cavity of the peritoneum, kidney tissue and blood.

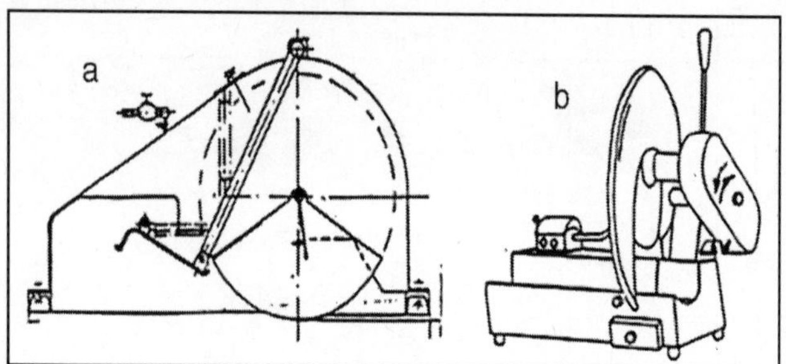

Fig. a. Deheading Machine with a Circular Saw, Operated Manually (Straight Cut), b. Deheading Machine with the Disc Saw and Guillotine Action (Straight Cut).

Fish is cut longitudinally up to the anal opening, and special care is taken to avoid cutting the gall bladder. This procedure is performed on a table made of special material which is hard, easy to wash and does not absorb fluids. The table surface should be frequently rinsed and periodically disinfected.

A specialised gutting work station, allows to safely cut fish down the belly (used mainly during processing of trout), remove the guts by vacuum suction and quickly wash and rinse the body cavity with a rotational brush and a water spray, including kidney tissue removal. Simple systems consisting of rotating brushes and water sprays are widely used. They facilitate the work and increase the product quality. Protective gloves, periodically disinfected and replaced, should be worn during gutting, especially when mechanised devices are used.

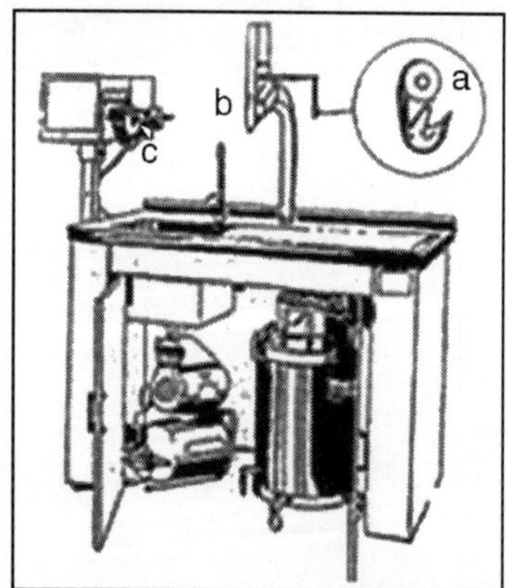

Fig. Gutting Work Station: a. Cutting Down Belly with a Safe Cutting Element, b. Removal of Guts with Vacuum Suction, c. Washing and Rinsing of Body Cavity with Rotational Brush.

It is likely that the vacuum suction tools (kidney and blood removal) used to clean the body cavity in processing salmonids, will find an application for other freshwater fish species.Gutting machines for processing trout, eel and a couple of other species, have been constructed in several countries, but high price renders them unsuitable for smaller plants. The cutting of the body cavity, removal of guts and kidney tissue with brushes and vacuum suction can be performed in these multi-application machines.

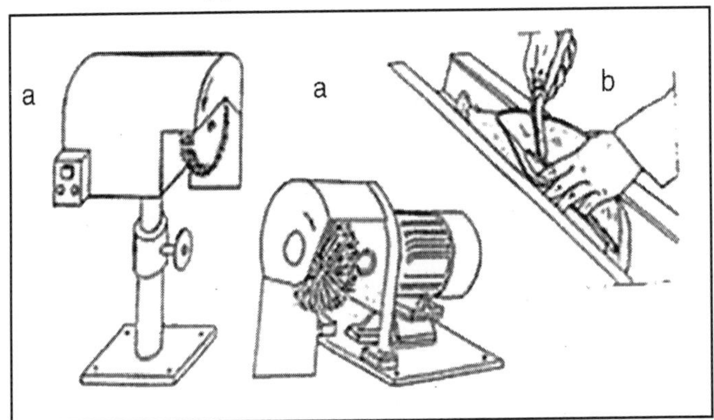

Fig. a. Rotating Brushers Used to Clean Body Cavity and to Remove Kidney Tissue (with and without Adjustable Height), b. Hand-held Vacuum Suction Tool for Kidney Removal and Cleaning of Body Cavity.

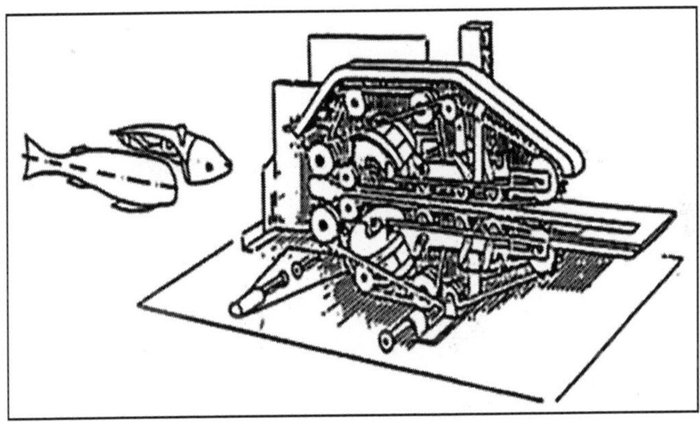

Fig. Deheading and Gutting Machine.

Some freshwater fish species, in particular bream, perch, roach, carp of length 20-40 cm, can be deheaded and gutted in a machine which employs a so-called American cut. Although the technological efficiency of this cut is not high, the processing speed reaches up to 40 fishes/minute.

Cutting Away the Fins

Manually cutting away the fins with either a knife, special mechanised scissors or rotating disc knives, is a labour-intensive and strenuous operation when handling larger fish. This operation is most frequently done after gutting during the production of deheaded whole fish and fish steaks. An automated device consisting of the rotating disc knives with a slit cutting edge, powered by electric motor, facilitates and speeds up the fin removal procedure. The knife slot has a horizontal opening through which the dorsal and ventral fins are passed manually and cut out.

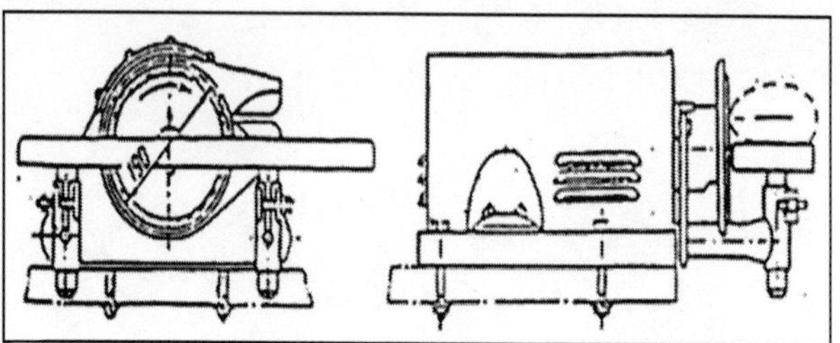

Fig. Equipment for Cutting out the Fins, Employing a Disc Knife.

Slicing of Whole Fish into Steaks

Slicing of deheaded whole fish into steaks with a cut perpendicular to the animal's backbone is a very common fish processing method. The high technological efficiency of this processing technique compared to filleting and automated cutting into pieces, makes it popular with retail markets and the canning industry. The fish pieces obtained average 2.5 to 4.5 cm thick.

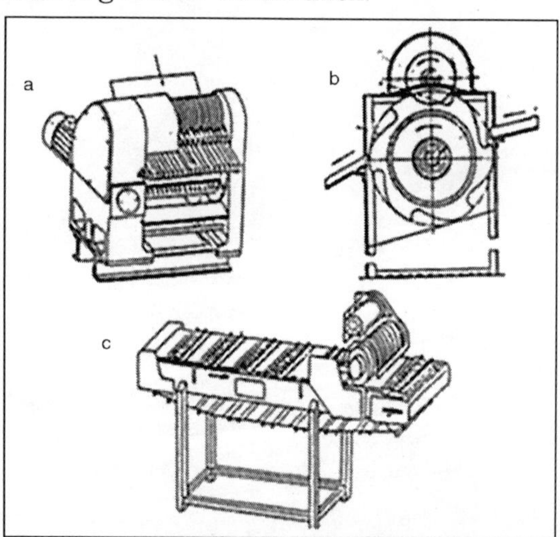

Fig. a. Cutter Used for Slicing Whole Fish into Steaks, b. Cutter with a Drum-type Loading System, c. Cutter with a Loading Conveyor Belt.

Smaller and medium size fish are cut manually in concave basins which have slots evenly spaced to facilitate slicing into steaks of equal thickness. A knife or a band saw is used to slice the fish. Sometimes a band saw is used to remove the head and cut the body into two parts, one retaining the backbone. Larger fish, particularly cyprinids, which have a massive and more solid backbone, need slicing mechanically. Numerous designs of such machines exist, and generally utilize multiple rotating circular saws attached to the drive. The distance between the saws as well as the elements moving the fish along the line can be adjusted. The deheaded whole fishes are placed into an automated cutter oriented so that the last piece cut has a prescribed length. A mechanised cutter can process 20-40 fishes/minute, depending on the fish size.

Filleting

A fillet which is a piece of meat consisting of the dorsal and abdominal muscles has been a most sought-after fish product in the retail market. Filleting efficiency depends upon fish species, its sex, size and nutritional condition. Manual filleting is very labour-intensive and largely depends on the skills of the workers. However, filleting of freshwater fish is not as widely applied as for marine fish.

Filleting machines for processing marine fish are quite costly and are not suitable for freshwater species; in the case of trout, for example, expensive multi-function devices have been designed which are not used in small processing plants. Some fish markets sell fillets of carp, perch, pike-perch and smoked single or block fillet of trout. Besides fillets, other forms are processed, *e.g.*, block fillet retaining some bones (boned fillet) and the simplest type of processed carp which is the deheaded whole fish cut into two halves, one retaining the backbone. Restaurants and fish stores use simple tools to streamline the manual longitudinal cutting of fish. The same result is obtained by using a filleting device with a single rotating disc knife and two conveyor belts.

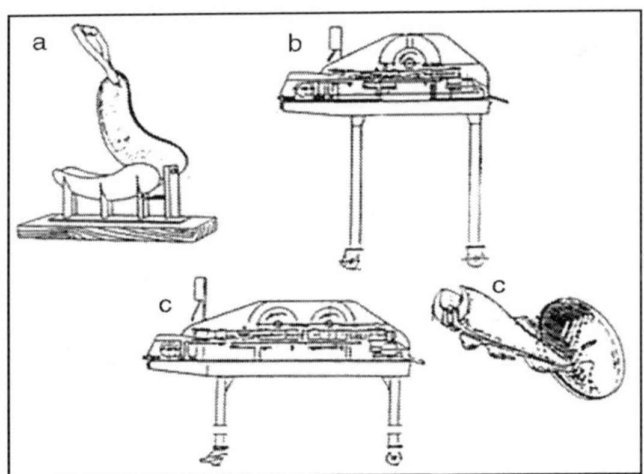

Fig. a. Manual Tool for Cutting Whole Fish into Two Parts (Backbone Remains), b. Filleting Machine with Conveyors, Used for Cutting Fish into Two Halves (Backbone Remains), c. Filleting Machine with a Conveyor, Used in Production of Boned Fillet without the Backbone.

Manual filleting and deboning are time- and labour-consuming procedures, and are usually carried out using simple and inexpensive machines. In small plants processing freshwater fish, a type of machine which separates fillets and bones, sometimes with part of the backbone left near the head region, is increasingly more common. The demand for freshwater fish fillets increases interest in simple and inexpensive single-purpose machines for filleting of deheaded and/or gutted fish. Different species (trout, perch, pike-perch, pike, cyprinids, etc.) can be processed in these devices as long as they are in the same size range. The remaining ribs and pin bones are manually removed from the fillets, and sometimes, as in case of cyprinids, perch and roach, the bones are cut by machine.

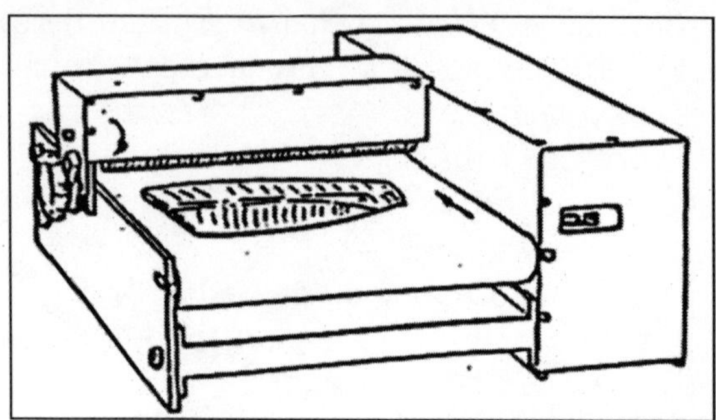

Fig. Machine Used for Cutting Ribs in the Boned Freshwater Fish Fillet.

The simplest filleting machine for gutted and deheaded fish has two disc knives set from each other at a distance equal to the thickness of the fish's backbone. Filleting speed of these devices is 30-40 fishes/minute: they are efficient and the quality of the final product is good. However, manual processing yields better results.

The size range of the processed fish is 20-45 cm. Machines of different design and with bigger knives are used for processing larger fish. Filleting devices are produced in several countries (Germany, Poland, Russia) and are increasingly used in small processing plants.

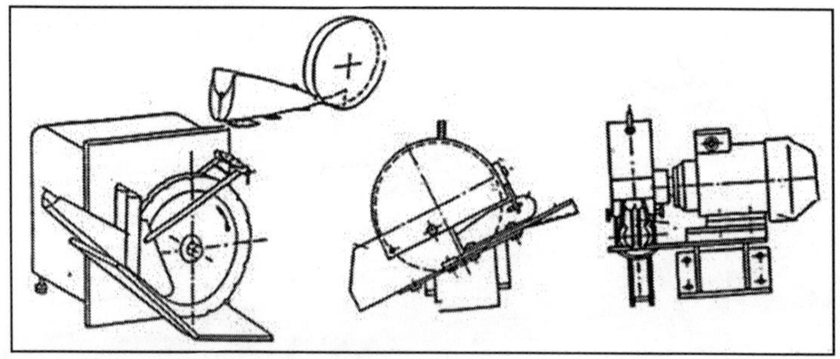

Fig. Filleting Machine Used for Production of Boned Fillets from Freshwater and Marine Fish.

Meat left on the fish's backbone after filleting can be recovered to a high degree using a meat-bone separator. Up to 50 per cent of the total mass of processed backbones can be recovered as meat. Boned fillets with ribs are subsequently processed by cutting the ribs in an automated system consisting of several disc knives 100-200 cm in diameter, set on a drive every 4-5 mm.

After cooking, particularly after frying, the tiny cut rib pieces are barely noticeable and cause no discomfort during consumption. In the machine used for cutting ribs, the boned fish fillets lie skin-down on a conveyor belt which drives them under the disc knives; the ribs are cut and incisions of determined depth are made in the meat.

Skinning

Only recently has skinning of freshwater fish fillets been introduced into processing plants. Manual fillet skinning is labour-intensive and difficult; a sharp knife and flat board made of metal or plastic are needed. The fillet is placed on the board skin-down, the meat is grasped in the left hand and the knife is drawn between the skin and meat. The simplest and most inexpensive automated tool for skinning of fillet with or without scales has been in use since 1992, and it can be attached to the processing table.

This tool consists of an oscillating knife powered with a small electric motor and a system of compression springs operated with a foot pedal. Water is not needed to operate this device. One end of the fillet is placed in a slit between the knife and compression element and the tip grasped manually in a wrench which allows the skin to be pulled off the meat from under the oscillating knife. Various freshwater and marine fish species can be processed in this machine, including larger fish.

Its use is recommended for small processing plants, fish markets, fishmongers, supermarkets, restaurants and catering sectors. Compared with manual operations, this machine facilitates and speeds up skinning. Some devices are small and can be placed directly on the processing table; running water and electricity are necessary for their operation. Efficiency varies depending upon the fish species.

The price of these devices varies; some are quite expensive and their use is profitable only when a certain level of production is maintained. Depending on fillet size and type of machine, 20 to over 40 fillets/minute can be skinned; faster machines require a conveyor to move the fillets. Skinning machines are produced in many countries.

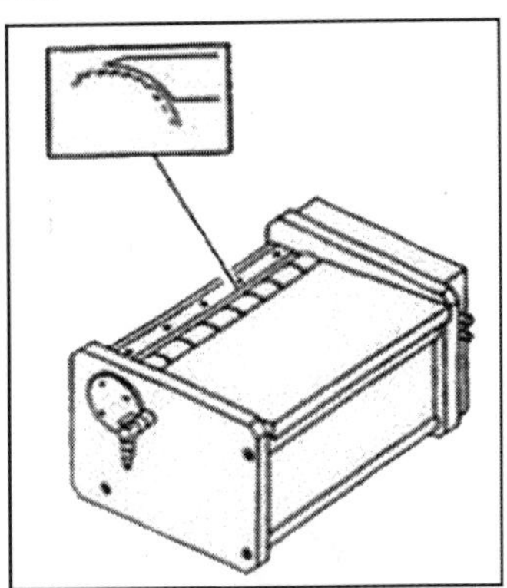

Fig. Skinning Machine with Stationary Knife.

Meat-Bone Separation

In recent years a new trend has emerged to effectively process raw fish products which resulted in production of minced meat separated from inedible parts, such as bones, skin and scales. During filleting a considerable amount of meat is usually left along the ribs and backbone (30-50 per cent). The carcasses are a source of minced meat. Minced meat is also produced from less valuable fish species after deheading, their body cavities carefully cleaned and kidney tissue removed. Meat is separated from the bones, skin and scales, in automated devices called separators. In the separator, meat is squeezed through holes into the cylinder under pressure applied by a conveyor belt partially encircling the cylinder (about 25 per cent of the cylinder's perimeter). The cylinder rotates slightly faster than the conveyor. The openings in the cylinder are usually 3-7 mm in diameter. For processing of freshwater fish, the holes are 4 and 5 mm in diameter. The smaller the holes, the stronger the grinding action. Pressure applied by the conveyor to the cylinder can be regulated depending on the type and size of the raw product and on the hole diameter.

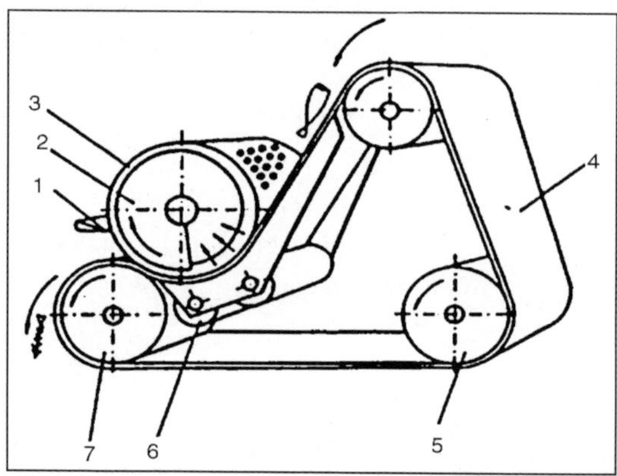

Fig. Diagram of Meat Separator: 1- Scraper, 2- Screw Type Meat Remover, 3- Perforated Drum, 4- Rubber Band, 5- Band Tension roller, 6- Pressing Roller, 7- Transmission Roller.

The use of separators for processing such freshwater species as perch, bream and tench, offers a new perspective on production of novelty products which could gain customer approval and be successfully marketed. Minced meat can be either frozen in cardboard or foil containers, or used immediately to produce fishburgers, fish sticks, canned fish, vegetable mixes and fish dumplings.

The technological efficiency attained during the production of ground meat from bream not larger than 1 kg, was 40 per cent of total body weight. For example, in Poland in a small fish processing plant which employs 8 workers, 1 t of frozen ground bream meat can be produced during one shift. According to routine practice, ground meat can be stored at -25°C to -28°C for up to6 months.

NUTRITIVE AND TECHNOLOGICAL VALUES OF FRESHWATER FISH

The manufacturing potential of the raw material as food depends on two features - the nutritive and the technological value.

Nutritive Value

The nutritive value of dishes prepared from fish and from animal meat is comparable, but in some cases fish-based meals are advisable. In such an evaluation, many parameters, such as energetic value, quality and content of protein components, vitamins and mineral compound content should be examined. The energetic value of eel meat is lower than that of fat beef (1 050 kJ/100 g and 1 250 kJ/100 g respectively); while in the case of trout it amounts to 600 kJ/100 g and it is lower than for lean beef, 735 kJ/100 g. Thus the meat of freshwater fish can be a valuable constituent in low-calorie diets and at the same time has a high energy content.

The composition of amino-acid proteins in fish meat is similar to that of a hen's egg. Consumption of fish together with products of plant origin which are poor in some amino-acids (lysin, threonine), enables not only a complete utilisation of plant protein, but also improves the content of a diet. The biological value of freshwater fish fats is lower than that of marine fish because the former contain fewer unsaturated aliphatic acids. Fish meat is valuable as a source of vitamins and mineral substances. It contains especially the trace metals such as: selenium, molybdenum, cobalt, whose value is emphasised by physiologists.

The definition of food stresses that the basic food ingredients as well as the raw materials used for its production must be wholesome. However, contamination of the environment is fast increasing, especially through the use of chemicals in agriculture or in industry. For that reason, certain countries or groups of countries establish limits and recommendations for permissible levels of chemical contaminants the excess of which leads to exclusion of such raw material from the production of food for human consumption. This problem, for many reasons (diets, habits, analytical methods), is far from being solved as countries have different attitudes in this respect. It may well constitute a non-tariff barrier on a free market in the future.

Technological Value

The technological value generally depends on two parameters: the yield of preliminary processing and the quality features of fish meat and by-products. The yield of edible parts of the fish depends, first, on the species and constitution, and also on age and consequently size and maturity. Yield is affected by the ratio between edible and inedible parts of the fish and this is a decisive factor with regard to the technological value of the fish. This ratio depends on the species.

It is most favourable in the Salmonidae family, amounting to approximately 75 per cent of the weight. For most fish species this parameter ranges from 50 to 60 per cent. In the case of perch and most of the Cyprinidae family the yield is less than 50 per cent. More information on the yield of preliminary processing of freshwater fish. Evaluation of the technological value of freshwater fish should take account of its possible utilisation for different products, considering the sensory properties such as: flavour, texture, appearance, size and bone content. These parameters are decisive as to consumer's interest and thus the market demand.

Table: Average yield of preliminary processing (manual processing) of several species of freshwater fish

Species	Size of Fish [Kg]	Form of Preprocessed Fish	Yield [%]
Trout	> 0.35	gutted	74 - 82
Trout	> 0.35	h/g (deheaded/gutted)	62 - 74
Trout	> 0.35	fillet with skin	50 - 55
Carp	> 3.0	gutted	76 - 82
Carp	> 1 - 3.0	gutted	73 - 79
Carp	> 3.0	deheaded and gutted	55 - 61
Carp	> 3.0	chunks	49 - 57
Carp	> 3.0	fillet with skin	41 - 49
Pike-perch	> 1.0	gutted	79 - 89
Pike-perch	> 1.0	deheaded and gutted	66 - 74
Pike-perch	0.35 - 0.5	deheaded and gutted	60 - 68
Pike-perch	> 1.0	chunks	56 - 68
Pike-perch	> 1.0	fillet with skin	52 - 64
Pike	1 - 3.0	gutted	76 - 84
Bream	> 1.0	gutted	68 - 76
Bream	0.5 - 1.0	deheaded and gutted	56 - 64
Bream	0.5 - 1.0	chunks	52 - 64

Fish with high bone content are not so popular as a product for consumption. Therefore, the technological value of roach (Rutilus rutilus) is lower than that of pike-perch (Stizostedion lucioperca). The taste of freshwater fish depends mainly on the quality of their water habitat and on their food. It is known that fish (for example, carp) living in dirty and muddy ponds, have an unpleasant flavour. The flavour of wild trout from streams is better than that of fish from aquaculture. The opposite is true for eel resulting from the fact that the aquaculture eel has a more tender tissue and thinner skin.

Freshwater fish are classified according to size, larger individuals usually being preferred to small fish. This is also connected with bone content: *e.g.*, trouts weighing about 300 g are very popular as single portions, prices increasing with popularity. The most expensive are fish weighing over 500 g which are destined for smoking. The best market value are carp of 1-2 kg, but those exceeding 3 kg have less customer appeal. The sanitary and hygienic condition of fish and fish meat also influences the technological value. This relates to the presence of parasites and pathogenic micro-organisms. However, the main role in evaluating technological value and usefulness is played by a set of features termed freshness. These features change during storage after the death of the fish and the intensity of the changes depends on the species, fishing conditions and storage conditions immediately after capture.

Post-Mortem Changes and Fish Quality Assurance Methods

On the death of the fish, processes of physical and chemical change caused by enzymes and micro-organisms begin to occur. The complete decay of the fish is the final result of those changes.

Post-mortem changes which take place in fish tissue occur in the following phases:

- Slime secretion on the surface of fish
- Rigour mortis
- Autolysis as enzymatic decomposition of tissues
- Microbiological spoilage

The duration of each phase can change or phases can overlap. This depends on storage conditions, especially the temperature which greatly influences these processes.

Slime Secretion

Slime is formed in certain cells of fish skin and the process becomes very active just after fish death. Some of the fish, for example eel, secrete more slime than, for comparison, Salmonidae and perch. Fish which secrete great quantities of slime have poorly developed scales; very often the quantity of slime reaches 2-3 per cent of the fish mass and that in turn creates problems during processing. The secretion process stops with the onset of rigour mortis. Slime contains large amounts of nitrogenous compounds and these provide good nourishment for micro-organisms originating from the environment. Therefore, the slime spoils quickly: first giving an unpleasant smell to the fish, and second opening the way for further and deeper bacterial penetration into the fish.

Rigour Mortis

Rigour mortis is a result of complicated biochemical reactions which cause muscle fibres to shorten and tighten, and finally the fish becomes stiff. Rigour mortis has many technological consequences. If, for example, the bones were removed prior to rigour mortis the length of the fillet shortens by 30 per cent. At the same time, the fillet becomes wider and thicker because its volume does not change. This tightness very often causes the connective tissue of individual myomeres to break; this process is termed "gaping" and results in muscle separation which is considered a quality defect. "Gaping" depends on temperature; the higher the temperature of fish at the beginning of the rigour mortis process the greater the gaping of the muscle. Therefore, during rigour mortis fish temperature should be as low as possible. For example, for roach and perch kept at 0° C rigour mortis begins 24 hours after death and lasts for 72-80 hours. When the same species is kept at 35° C it begins 20-30 minutes after death and stops after about 3 hours. The time rigour mortis begins and its duration depend on the fish species (*e.g.*, for carp at 0° C it starts after 48 hours, for roach and perch at 0° C after 24 hours), on the fish catching technique, and on fish temperature. It was also found that fast swimmers, for example trout, undergo rigour mortis faster but for a shorter duration than slow swimmers like carp.

In those fish which are in good condition (well-nourished) rigour mortis is more intensive. Fish put to death just after removal from the water reach a state of rigour mortis later than those fish which died after a long agony. In the case of carp put to death just after capture rigour mortis begins after 48 hours, but if the carp died after a long agony it sets in after 24 hours (at 0° C).

Unnecessary and rough handling of the fish can shorten the time of occurrence and duration of rigour mortis. Such treatment causes stress in live fish. Fish body temperature is a decisive factor in the onset and duration of the rigour mortis process. The higher the temperature the sooner it begins and the faster it ceases. This is evidenced by enzymatic reactions whose speed increases with increased temperature. At high temperatures it results in greater changes in proteins, the latter causing higher loss of tissue juices, *e.g.*, during processing. Usually, the later rigour mortis begins and the longer it lasts, the longer are the storage life of the fish and its use for consumption.

Autolysis

On the death of the fish, a complicated biochemical process starts, leading to a decomposition of basic compounds of tissues which takes place under the influence of enzymes. This decomposition involves proteins, lipids and carbohydrates. Its intensity is not the same for all compounds and the decomposition of one can influence the decomposition of the others.

The quality of fish as a raw material for consumption or for processing depends largely on proteolysis, that is, the decomposition of proteins. This process follows rigour mortis. The final products of protein hydrolysis, under the influence of enzymes, are: amino-acids and other low-molecular substances which have an impact on the sensory features of fish. A similar situation concerns the products of lipid autolysis: thus autolysis cannot be qualified as a phase in the spoilage process.

During autolysis, great changes occur in the structure of muscle tissue which becomes softer, and very often falls into layers along the myosepts. In small fish, perforation of the belly occurs. From the technological view, it is negative because the proteolysis process leads to a decrease in the capacity of tissue to retain tissue juice, resulting in toughness of texture of the final product. The degradation of proteins creates ideal conditions for the growth of spoilage bacteria.

Microbiological Decomposition

The muscle tissue of live fish is generally sterile but bacteria thrive in the alimentary tract and on the skin, and from there they penetrate into the muscles; for example, through the blood vessels. This process is further favoured by structural changes in the tissue as a result of rigour mortis and autolysis. Bacteria are able to decompose proteins, but products of proteolysis such as amino-acids and other low-molecular nitrogenous compounds provide better nourishment.

Thus it was found that, due to lower content of these substances, freshwater fish tissue undergoes microbiological decomposition more slowly than marine fish tissue. Micro-organisms cause decomposition of not only proteins but other compounds containing nitrogen, lipids to peroxides, aldehydes, ketones and lower aliphatic acids. However, the decomposition of nitrogenous compounds occurs much faster than in the case of lipids.

Compounds such ammonia, hydrogen sulphide and mercaptans, indole, skatole, etc., are the final products of microbiological spoilage of fish, which produces an unpleasant and then disgusting flavour. Penetration of bacteria into fish tissue and microbiological decomposition begins with autolysis and these processes are practically parallel. However, their rate and intensity strictly depend on the storage temperature. Low temperature strongly inhibits the activity of micro-organisms in which case the autolysis process dominates.

INDICATORS OF FISH FRESHNESS

Freshwater fish, as other fish species, are raw material which fast deteriorates. This implies that both the producer and the consumer are very often exposed to the risk of buying fish which is not fresh or has even deteriorated. Knowledge of the average shelf life for individual fish species - depending on storage conditions - is a basic principle applied in the food - and the fish - industry. Effective, objective and repeatable methods for evaluation of raw material freshness should be specified, but attempts so far are only now showing positive results.

Thus, sensory analysis is the main method of evaluating fish freshness. It enables differences in texture, flavour, and taste to be determined, and subsequently the usefulness of the raw material. Sensory properties change during storage from the desired very high standard, through neutral or average, and finally to undesirable or disgusting. It is generally assumed that prior to disappearance of desirable features the fish is considered to be fresh, while the appearance of undesirable or disgusting features disqualifies the raw material. The most difficult step is to determine an intermediate state in which the fish is not entirely fresh. Sensory analysis is thus carried out on raw fish and cooked fish. Flavour, appearance and state of abdominal cavity (for not eviscerated fish) are the main indicators of quality in the case of raw fish. For cooked fish, smell is the most important indicator.

8

Seafood Modified Atmosphere Packing

Modified atmosphere packing, MAP, means replacing the air in a pack of fish with a different mixture of gases, typically some combination of carbon dioxide, nitrogen and oxygen. The proportion of each component gas is fixed when the mixture is introduced, but no further control is exercised during storage, and the composition of the mixture may slowly change. Modified atmosphere packing is often incorrectly called controlled atmosphere packing, CAP. Controlled atmosphere packing means packing in an atmosphere whose composition is continuously controlled throughout storage; such control is possible in large storage units, but not in small packs. The storage life of sonic chilled products, notably white fish, can be extended by packing in a modified atmosphere. The appearance of the pack is attractive and, since the transparent packaging is not in close contact with the contents, the buyer can clearly see the product. Modified atmosphere packs have the advantages common to most forms of prepacked fish; they are odourless, easy to label and convenient to handle. In addition they are leakproof and robust.

Modified atmosphere packing is relatively expensive, currently about twice the cost of vacuum packing. Continuous production of rigid packs entails the purchase of expensive packaging machinery and the use of expensive thermoformable film.Modified atmosphere packs are commonly two or three times bulkier than other forms of pack, and therefore are costlier to carry and store. The walls of a pack may collapse when the enclosed atmosphere contains a high proportion of carbon dioxide, which is highly soluble in fish tissue. As the carbon dioxide dissolves, a partial vacuum is created, the pack may collapse onto the product, and in some instances the contents may be squashed. The problem can be avoided by correct choice of gas mixture. Unsightly drip may form inside the pack when too high a proportion of carbon dioxide is used. The problem can be minimized by choosing the right gas mixture and by introducing an absorbent paper pad beneath the fish. Any extension of storage life attributable to modified atmosphere packing may be lost if the additional safeguard of chilled storage is ignored; the packs must be kept at or close to 0°C throughout distribution if the full benefits of a modified atmosphere are to be maintained.

Gas Mixture

Carbon dioxide retards bacterial spoilage of fish, but too high a proportion in the mixture can induce pack collapse, excessive drip and, in products that are eaten without further cooking, for example brown crab meat, an acidic, sherbet-like flavour. Oxygen can prevent colour changes and bleaching that would otherwise occur in some products. Nitrogen is an inert gas that is used to dilute the mixture. The following mixtures are recommended, based on an assumed ratio of 3 parts gas mixture to 1 part fish by volume in the pack.

For white fish, scampi, shrimp and scallops a mixture of 40 per cent carbon dioxide, 30 per cent nitrogen and 30 per cent oxygen gives the best results. For salmon, trout, fatty fish such as herring and mackerel, and for smoked fish products, a mixture of 60 per cent carbon dioxide and 40 per cent nitrogen is recommended. Smoked salmon packed in this mixture may sometimes show a green discolouration during storage, the extent of the greening being dependent on the strength of the smoke cure; where greening is likely to occur, the mixture recommended for white fish should be used.

Machinery and Materials

The simpler and cheaper machines pack the product in a flexible bag or on a tray inside a bag to form what is termed a pillow pack. At the more expensive end of the available range are sophisticated machines that continuously form packs with rigid bases from rolls of thermoformable plastics film. Some of these machines have a dual purpose and can be converted to vacuum packing when required, although the changeover can take up to 3 hours. Modified atmosphere packs can be produced at a rate of more than thirty a minute on the fastest machines. Packaging films must have a low gas permeability, and be to the machine maker's specification.

Gases can be obtained either ready mixed, or separately for use in machines that mix the gases before packing. Where gases are mixed in the machine, the gas composition in the packs should be measured at the start of a run, and monitored throughout the day, particularly when faults are suspected or adjustments to the gas mix are made.

Quality of Fish

Only the highest quality fish should be used for modified atmosphere packs, in order to gain the most benefit from any extension of storage life; packing fish in a modified atmosphere is not a means of marketing medium quality or poor quality fish. White fish should be of a quality equivalent to 1-4 days in ice, and should be free from blemishes and visible parasites. Herring and mackerel should be of a quality equivalent to 1-3 days in ice, and should contain at least 8 per cent fat. Smoked fish products should be made from fish of the same initially high quality. Salmon and trout should be of a quality equivalent to 1-3 days in ice.

PACKING THE FISH

Fish should be handled hygienically and kept chilled from the time of capture or harvesting until they are packed; whole fish and fillets should be kept in ice while awaiting processing, and smoked products should be held in a chillroom at 0°C. Ideally an air blast chiller should be provided in the processing line, either before or after the packing machine, since the fish may warm significantly during the packing operation.

Layering of products within a pack should be avoided; a single fillet or portion is more fully exposed to the action of the gases. Layering is unavoidable when packing sliced smoked salmon, but the product does not gain the full benefit of the modified atmosphere. Wet fish products that are likely to exude drip can be laid on a pad of absorbent paper inside the pack. Packs with faulty seals can be detected by pressing them with the hands; faulty packs will collapse. Packs should be clearly labelled according to existing regulations, and should be marked with a sell-by or consume-by date.

Storage Life of Packs

Storage life will depend on the species of fish used, its initial quality and fat content, the nature of the finished product, temperature of storage and, in a modified atmosphere, the gas mixture. Temperature of storage is of paramount importance in deriving the most benefit from a modified atmosphere; packs should be stored at a temperature as near to 0°C as possible, and never above 5°C. Any benefit from a modified atmosphere will be much reduced when storage temperature rises above 5°C.

White fish fillets benefit most from packing in a modified atmosphere; for example cod fillets of high initial quality, packed in the recommended gas mixture, and with a ratio of gas to product in the pack of 3:1 by volume, will keep up to 50 per cent longer at 0°C than when stored under vacuum or unwrapped.

Raw shell-on scampi and shrimp keep up to 30 per cent longer at 0°C in a modified atmosphere pack than in other types of pack, and the onset of black spot is inhibited. The storage life of herring, mackerel, salmon, trout, and smoked fish products is not extended in a modified atmosphere at 0°C.

Safety

Fishery products in the UK have a good record of safety with regard to food poisoning, and products in modified atmosphere packs are no exception. Some concern has been expressed about smoked products that are eaten without further cooking, for example smoked salmon and smoked mackerel, but the risk of an outbreak of botulism or of scombrotoxin poisoning from these products is no greater when packed in the recommended modified atmosphere than when packed in any other form.

PACKAGING USED IN FRESHWATER FISH PROCESSING

THE ROLE OF PACKAGING

The previous section discussed the processing methods most often used by small freshwater fish processing plants. Quality assurance is essential in each technological process, and suitable packaging materials and methods are of great importance. If these requirements are not met all efforts made during processing could be of little avail, which could lead to serious economic losses.

Packaging should protect the product from contamination and prevent it from spoilage, and at the same time it should:

- Extend shelf life of a product
- Facilitate distribution and display
- Give the product greater consumer appeal
- Facilitate the display of information on the product

The quality of freshwater fish which is delivered to the consumer or the processing plant as live fish greatly depends on correct handling during transport and, when processed, on suitable packaging. For short distances, the live fish can be transported in insulated containers with lids, capacity varying from 300 to 1 000 kg of fish. Fish can also be transported in normal lorries, but for long distances the water in the containers must be aerated and cooled by portable devices. In order to maintain good quality of fresh fish during transportation, fish boxes made of suitable materials should be used.

When purchasing fish boxes the six following requirements should be remembered; they should:

- Be of a suitable size for the range of fish to be handled or the product to be put into them
- Be of a convenient size for manual handling or lifting by mechanical equipment
- Be stackable such that the weight of the containers on top rests on the containers underneath and not on the fish
- Be constructed of impervious non-staining materials
- Be easy to clean
- Provide drainage for melted ice

Fish boxes are usually made of high-density polyethylene. Although this offers many advantages, such as duration, lightness, ease of cleaning, there are also disadvantages, *e.g.*, high price and the fact that they are not returnable. That is why disposable fish boxes of about 25 kg capacity (fish and ice) are more often used: these include fibreboard cartons, waxed and waterproof boxes. In the case of transport by lorries with no cooling system, insulated cartons, *e.g.*, boards made of moulded polyestyrene should be preferred. The latter is commonly used for delivery of chilled and frozen fish and fish products to wholesale and retail outlets. In the case of fillets, each layer of fillets should be packed thin and separated from the ice with a plastic foil. Styropor boxes are normally sold with lids, which fit very closely and can be with or without drainage holes. In a typical range, wall thickness varies with box size; *e.g.*, a 6 kg capacity box has a 15 mm thick wall, a 10 kg box a 19 mm wall, a 25 kg box a 25 mm wall. The main disadvantage of moulded polyestyrene fish boxes is their lack of strength. They are easily damaged or broken by rough handling. This limits their size and use.

Polyestyrene is difficult to clean. Polyestyrene boxes are difficult to re-use, and are usually non-returnable. They may cause disposal problems due to their bulk. The packaging industry improves its products by using new materials with better insulating properties or by introducing new leakproof designs. The new containers are often lighter and less bulky. For example, the Therma Gard packing system consists of a metallised plastic bag (which reflects practically all radiant heat). This is then wrapped in a waterproof and leakproof carton. The metallised bag, together with a bubble-pack wrapper, provides a double-pack insulation. The Therma Gard bag can be sealed airtight and thus be used for carrying live fish. The Stratech aluminised boxes have a wall thickness of only 5mm and it is claimed that these boxes have similar insulating characteristics as polystyrene boxes with 30 mm wall thickness.

The future use of expendable packages is becoming questionable as there is a growing discussion, for example in some states of the USA, on imposing a ban on these packages. The main drawbacks in using returnable containers are freight costs for returning empty containers. Use of "knock-down" returnable containers will reduce freight costs.

RETAIL PACKAGING FOR FRESHWATER FISH PRODUCTS

The main role of packaging is described above but in respect of retail presentation it should also reduce the smell and the drip, and enable the product to be tucked into shopping baskets with other purchases. Moreover, the packaging of fish products should ensure attractive presentation among other food products without contaminating them. Basic packaging materials include paper, cartons, sheets of metal, metal foils and many kinds of plastics. Despite the rapid growth in use of plastics, the role of paper and carton as packaging materials does not decrease.

Kraft paper or carton are often laminated with polyethylene or aluminium foil which render them waterproof. Such material is used for production of trays for packaging of fresh or frozen products. More often, trays are made of plastic materials such as polyestyrene or expanded polyestyrene. Expanded polyestyrene is frequently used but it is partly oxygen-permeable and so those products which are sensitive to rancidity have to be additionally overwrapped or skin-packed with suitable film. The materials mentioned above are not stable at high temperatures and hence are not suitable for trays to be used in an oven.

Polyester can be used as a packing material for heating of the product in the traditional and microwave ovens, but this material cannot be used for microwave cooking. Trays used for packing are generally overwrapped with a protective film, often with PE wrapping which shrinks. The film shrinking is achieved by use of hot air or hot water.

Stretch wrapping is often used for products which are heat-sensitive. The film is stretched over the product manually (very often in the supermarket) or by machine. Foils used as wrapping or bags for packing of trays with product must be puncture-proof, extensible and impervious to gases like oxygen. Hundreds of different films are used in the packaging industry.

These can be broadly categorised into two groups:

1. Basic films consisting of a single layer of film
2. Laminate consisting of two or more basic films glued together or bonded together by heat or by adhesives

Plastics such as polyethylene film or copolymer of ethylene and vinyl acetate are very often used for packing of frozen products. Polyethylene packs can be produced manually using pre-made bags. An impulse or bar sealer is used to seal the bags which are hand-filled.

In order to improve the barrier properties of packages laminates are used, for example polyester/polythene. Products which are particularly sensitive to oxygen are vacuum-packed. During the sealing operation, air is removed from the package. A laminate nylon/polythene is commonly used as packaging material. This type of packaging is used, for instance, for smoked trout which are arranged on a board with, for example, a coated texture. Numerous machines exist for vacuum-packing with single, double or continuous chambers. Vacuum-sealing machines can additionally be equipped with a modified atmosphere packing system (MAP).

Immediately on removing the air from the package a mixture of gases is pumped in. Usually this mixture consists of 30 per cent nitrogen, 40 per cent carbon dioxide, and 30 per cent oxygen. In the case of fat fish the oxygen is replaced by nitrogen. This method is increasingly used for packing fresh fish. The MAP products have to be stored at the temperatures lower than 3° C because of C. botulinum hazard. MAP packages consist of two kinds of foil. The bottom film is foil-rigid or semi-rigid. This foil is formed by, for example, extrusion and the resultant tray is moved to the packing section. Because of product drip it is placed on an absorbing board. The top web is drawn over the filled trays and sealed round the edges. The pack may be evacuated or gas-flushed before sealing.

Vacuum-skin packaging is becoming more common for packing smoked fish. In this process the wrapper is heated and wrapped over the product, the film moulding completely to the product shape and sealing the product completely, forming an extra skin.

Labelling Requirements for Freshwater Fish Products

Lack of detailed standards and existence of only limited regulations concerning wholesomeness and sanitary conditions for production and trade of food products characterise the market economy. Here, the problem of labelling is of a particular importance. Regulations in this regard are very detailed and are aimed at protecting the health of the consumer and providing the best information. These requirements enable the consumer to decide which products to buy. A label placed on the product should inform the consumer about the raw material used, method of preparation and form of consumption, shelf life, etc.

Product labelling is of prime concern in the European Union. Directive 79/112/EEC of 18 December 1978 was revised several times, and in 1990 there came into force a new Directive 90/496/EEC which concerned labelling and providing information on nutritive and energetic values (kcal or kJ/100 g or 100 ml), the amount of basic ingredients and nutritive compounds such as: proteins, carbohydrates, fat, fibre, sodium and vitamin content (EEC, 1979). These requirements were supplemented in Directive 89/396/EEC recommending the labelling of batches of product which would make it easier to withdraw the batch from commodity turnover in the case of health hazard. Taking into account the necessity to ensure complete information on the product to facilitate the selection of a healthy and economic diet, the US Food and Drug Administration (FDA) recently proposed a voluntary Nutritive Labelling Programme which covers, *inter alia*, a proposal for placing on the product for example information concerning the percentage of recommended daily intake of protein, vitamin A and C, iron, calcium, etc.

VACUUM AND MODIFIED ATMOSPHERE PACKAGED FISH AND FISHERY PRODUCTS

POTENTIAL FOOD SAFETY HAZARD

Clostridium botulinum toxin formation can result in consumer illness and death. This chapter covers the potential for C. botulinum growth and toxin formation as a result of time/temperature abuse during processing, storage and distribution. When C. botulinum grows it can produce a potent toxin, which can cause death by preventing breathing. It is one of the most poisonous naturally occurring substances known. The toxin can be destroyed by heat (*e.g.*, boiling for 10 minutes), but processors cannot rely on this as a means of control.

There are two major groups of C. botulinum, the proteolytic group (*i.e.*, those that break down proteins) and the non-proteolytic group (*i.e.*, those that do not break down proteins). The proteolytic group includes C. botulinum type A and some of types B and F. The non-proteolytic group includes C. botulinum type E and some of types B and F. The vegetative cells of all types are easily killed by heat. C. botulinum is able to produce spores. In this state the pathogen is very resistant to heat. The spores of the proteolytic group are much more resistant to heat than are those of the non-proteolytic group. The conditions under which the spores of the most heat resistant form of non-proteolytic C. botulinum, type B, are killed. However, there are some indications that substances that may be naturally present in some products, such as lysozyme, may enable non-proteolytic C. botulinum to more easily recover after heat damage, resulting in the need for a considerably more aggressive process to ensure destruction.

Temperature abuse occurs when product is exposed to temperatures favourable for C. botulinum growth for sufficient time to result in toxin formation. The conditions under which C. botulinum and other pathogens are able to grow. Packaging conditions that reduce the amount of oxygen present in the package (*e.g.*, vacuum packaging) extend the shelf life of product by inhibiting the growth of aerobic spoilage bacteria. The safety concern with these products is the increased potential for the formation of C. botulinum toxin before spoilage makes the product unacceptable to consumers.

There are a number of conditions that can result in the creation of a reduced oxygen packaging environment.

They include:

- Vacuum packaging or modified or controlled atmosphere packaging. These packaging methods directly reduce the amount of oxygen in the package;
- Packaging in hermetically sealed containers (*e.g.*, double seamed cans, glass jars with sealed lids, heat sealed plastic containers), or packing in deep containers frosm which the air is expressed (*e.g.*, caviar in large containers), or packing in oil. These and similar processing/ packaging techniques prevent the entry of oxygen into the container. Any oxygen present at the time of packaging may be rapidly depleted by the activity of spoilage bacteria, resulting in the formation of a reduced oxygen environment.

Packaging that provides an oxygen transmission rate of 10,000 cc/m2/24hrs (*e.g.*, 1.5 mil polyethylene) can be regarded as an oxygen-permeable packaging material for fishery products. This can be compared to an oxygen-impermeable package which might have an oxygen transmission rate as low as or lower than 100 cc/m2/24hr (*e.g.*, 2 mil polyester). An oxygen permeable package should provide sufficient exchange of oxygen to allow aerobic spoilage organisms to grow and spoil the product before toxin is produced under moderate abuse temperatures. However, use of an oxygen permeable package will not compensate for the restriction to oxygen exchange created by practices such as packing in oil or in deep containers from which the air is expressed. C. botulinum forms toxin more rapidly at higher temperatures than at lower temperatures. The minimum temperature for growth and toxin formation by

C. botulinum type E and non-proteolytic types B and F is 38°F (3.3°C). For type A and proteolytic types B and F, the minimum temperature for growth is 50°F (10°C). As the shelf life of refrigerated foods is increased, more time is available for C. botulinum growth and toxin formation. As storage temperatures increase, the time required for toxin formation is significantly shortened. Processors should expect that at some point during storage, distribution, display or consumer handling of refrigerated foods, proper refrigeration temperatures will not be maintained (especially for the non-proteolytic group). Surveys of retail display cases indicate that temperatures of 45-50°F (7-10°C) are not uncommon. Surveys of home refrigerators indicate that temperatures can exceed 50°F (10°C) (FDA, 2001a).

Control Measures

In reduced oxygen packaged products in which the spores of non-proteolytic C. botulinum are inhibited or destroyed (*e.g.*, smoked fish, pasteurised crabmeat, pasteurised surimi), normal refrigeration temperatures of 40°F (4.4°C) are appropriate because they will limit the growth of proteolytic C. botulinum and other pathogens that may be present. Even in products where non-proteolytic C. botulinum is the target organism for the pasteurisation process and vegetative pathogens, such as Listeria monocytogenes, are not likely to be present (*e.g.*, pasteurised crabmeat, pasteurised surimi), a storage temperature of 40°F (4.4°C) is still appropriate because of the potential survival through the pasteurisation process and recovery of spores of non-proteolytic C. botulinum aided by naturally occurring substances, such as lysozyme. In this case refrigeration serves as a prudent second barrier. In reduced oxygen packaged products in which refrigeration is the sole barrier to outgrowth of non-proteolyticC. botulinum and the spores have not been destroyed (*e.g.*, vacuum packaged raw fish, unpasteurised crayfish meat), the temperature must be maintained at 38°F (3.3°C) or below from packing to consumption. Ordinarily processors can ensure that temperatures are maintained at or below 38°F (3.3°C) while the product is in their control. However, current distribution channels do not ensure the maintenance of these temperatures after the product leaves their control. The use of time temperature integrators on each consumer package may be an appropriate means of enabling temperature control throughout distribution. Alternatively, products of this type may be safely marketed frozen, with appropriate labeling. For some products, control of C. botulinum can be achieved by breaking the vacuum seal before the product leaves the processor's control.

FDA Guidelines

Refrigerated, reduced oxygen packaged smoked and smoke-flavoured fish:

- For cold-smoked fish, the smoker temperature must not exceed 90°F (32.2°C).

- For hot-smoked fish, the internal temperature of the fish must be maintained at or above 145°F (62.8°C) throughout the fish for at least 30 minutes.
- Smoked fish must have not less than 3.5 percent water phase salt, or, where permitted, the combination of 3.0 percent water phase salt and not less than 100 ppm nitrite.
- Strict refrigeration control (*i.e.*, at or below 40°F [4.4°C]) during storage and distribution must be maintained to prevent growth and toxin formation by C. botulinumtype A and proteolytic types B and F and other pathogens that may be present.

Refrigerated, reduced oxygen packaged, pasteurised fishery products, which are pasteurised in the final container:

- Strict refrigeration control (*i.e.*, at or below 40°F [4.4°C]) must be maintained during storage and distribution to prevent growth and toxin formation by C. botulinum type A and proteolytic types B and F, and because of the potential survival through the pasteurisation process and recovery of spores of non-proteolytic C. botulinum aided by naturally occurring substances, such as lysozyme.

Refrigerated, reduced oxygen packaged, pasteurised fishery products, which are hot filled into the final container:

- Product temperature must be 185°F (85°C) or higher as the product enters the final container.
- Containers must be sealed according to the container or sealing machine manufacturer's seal guidelines.
- There must be a measurable residual of chlorine, or other approved water treatment chemical, at the discharge point of the container cooling tank; or the processor must follow the equipment manufacturer's UV light intensity and flow rate guidelines.
- Strict refrigeration control (*i.e.*, at or below 40°F [4.4°C]) must be maintained during storage and distribution to prevent growth and toxin formation by C. botulinum type A and proteolytic types B and F, and because of the potential survival through the pasteurisation process and recovery of spores of non-proteolytic C. botulinum aided by naturally occurring substances, such as lysozyme.

Refrigerated, reduced oxygen packaged "pickled" fish, caviar, and similar products:

- Growth and toxin formation by C. botulinum type E and non-proteolytic types B and F must be controlled by either:
  - Adding sufficient salt to produce a water phase salt level of at least 5 percent;
  - Adding sufficient acid to reduce the acidity (pH) to 5.0 or below;

- Reducing the amount of moisture that is available for growth (water activity) to below 0.97; or
- Making a combination of salt, pH, and/or water activity adjustments that, when combined, prevent the growth of C. botulinum type E and non-proteolytic types B and F.

- Processors should ordinarily restrict brining and pickling loads to single species and to fish portions of approximately uniform size. This minimizes the complexity of controlling the operation.
- Strict refrigeration control (*i.e.*, at or below 40°F [4.4°C]) must be maintained during storage and distribution to prevent growth and toxin formation by C. botulinum type A and proteolytic types B and F.

Refrigerated, reduced oxygen packaged raw, unpreserved fish and unpasteurised, cooked fishery products:

- For refrigerated, reduced oxygen packaged raw, unpreserved fish (*e.g.*, vacuum packaged fresh fish fillets) and unpasteurised, cooked fishery products (*e.g.*, vacuum packaged, unpasteurised crabmeat, lobster meat, or crayfish meat), the sole barrier to toxin formation by C. botulinum type E and non-proteolytic types B and F during finished product storage and distribution is refrigeration. These types of C. botulinum will grow at temperatures as low as 38°F (3.3C). As was previously stated, maintenance of temperatures at or below 38°F (3.3°C) after the product leaves the processor's control cannot normally be ensured. Time temperature integrators on each consumer package may be an appropriate means of providing such control.

Frozen, reduced oxygen packaged fishery products:

- If your product is immediately frozen after processing, maintained frozen throughout distribution, and labeled to be held frozen and to be thawed under refrigeration immediately before use (*e.g.*, "Important, keep frozen until used, thaw under refrigeration immediately before use"), then formation of C. botulinum toxin may not be a significant hazard.

Unrefrigerated (shelf-stable), reduced oxygen packaged fishery products:

- Spores of Clostridium botulinum types A, B, E and F must be destroyed after the product is placed in the finished product container or a barrier, or combination of barriers, must be in place that will prevent growth and toxin formation by Clostridium botulinum types A, B, E and F, and other pathogens that may be present in the product. Suitable barriers include:
  - Sufficient salt is added to produce a water phase salt level (the concentration of salt in the water-portion of the fish flesh) of at least 20 percent (Note: this value is based on the maximum salt level for growth of S. aureus.)

- Sufficient salt is added to reduce the water activity to 0.85 or below;
- Sufficient acid is added to reduce the pH to 4.6 or below;
- The product is dried sufficiently to reduce the water activity to 0.85 or below.

FOOD CODE GUIDELINES

Reduced Oxygen Packaging (ROP)

ROP which provides an environment that contains little or no oxygen, offers unique advantages and opportunities for the food industry but also raises many microbiological concerns. Products packaged using ROP may be produced safely if proper controls are in effect. Producing and distributing these products with a HACCP approach offer an effective, rational, and systematic method for the assurance of food safety. The purpose of this Annex is to provide guidelines for effective food safety controls for retail food establishments covering the receipt, processing, packaging, holding, displaying, and labeling of food in reduced oxygen packages.

Definitions

The term ROP is defined as any packaging procedure that results in a reduced oxygen level in a sealed package.

The term is often used because it is an inclusive term and can include other packaging options such as:

- Cook-chill is a process that uses a plastic bag filled with hot cooked food from which air has been expelled and which is closed with a plastic or metal crimp.
- Controlled Atmosphere Packaging (CAP) is an active system which continuously maintains the desired atmosphere within a package throughout the shelf-life of a product by the use of agents to bind or scavenge oxygen or a sachet containing compounds to emit a gas. Controlled Atmosphere Packaging (CAP) is defined as packaging of a product in a modified atmosphere followed by maintaining subsequent control of that atmosphere.
- Modified Atmosphere Packaging (MAP) is a process that employs a gas flushing and sealing process or reduction of oxygen through respiration of vegetables or microbial action. Modified Atmosphere Packaging (MAP) is defined as packaging of a product in an atmosphere which has had a one-time modification of gaseous composition so that it is different from that of air, which normally contains 78.08 per cent nitrogen, 20.96 per cent oxygen, 0.03 per cent carbon dioxide.
- Sous Vide is a specialised process of ROP for partially cooked

ingredients alone or combined with raw foods that require refrigeration or frozen storage until the package is thoroughly heated immediately before service. The sous vide process is a pasteurisation step that reduces bacterial load but is not sufficient to make the food shelf-stable. The process involves the following steps:

- Preparation of the raw materials (this step may include partial cooking of some or all ingredients);
- Packaging of the product, application of vacuum, and sealing of the package;
- Pasteurisation of the product for a specified and monitored time/ temperature;
- Rapid and monitored cooling of the product at or below 3°C(38°F) or frozen; and
- Reheating of the packages to a specified temperature before opening and service.
- Vacuum Packaging reduces the amount of air from a package and hermetically seals the package so that a near-perfect vacuum remains inside. A common variation of the process is Vacuum Skin Packaging (VSP). A highly flexible plastic barrier is used by this technology that allows the package to mold itself to the contours of the food being packaged.

Benefits of ROP

ROP can create a significantly anaerobic environment that prevents the growth of aerobic spoilage organisms, which generally are Gram negative bacteria such as Pseudomonads or aerobic yeast and molds. These organisms are responsible for off-odors, slime, and texture changes, which are signs of spoilage. ROP can be used to prevent degradation or oxidative processes in food products. Reducing the oxygen in and around a food retards the amount of oxidative rancidity in fats and oils. ROP also prevents colour deterioration in raw meats caused by oxygen. An additional effect of sealing food in ROP is the reduction of product shrinkage by preventing water loss.

These benefits of ROP allow an extended shelf-life for foods in the distribution chain, providing additional time to reach new geographic markets or longer display at retail. Providing an extended shelf-life for ready-to-eat convenience foods and advertising foods as "Fresh-Never Frozen" are examples of economic and quality advantages.

Safety Concerns

Use of ROP with some foods can markedly increase safety concerns. Unless potentially hazardous foods are protected inherently, simply placing them in ROP without regard to microbial growth will increase the risk of foodborne illnesses. ROP processors and regulators must assume that during distribution of foods or while they are held by retailers or consumers, refrigerated temperatures may not be consistently maintained. In fact, a serious concern is that the increased use of vacuum packaging at retail supermarket deli-type operations may be followed by temperature abuse in the establishment or by the consumer. Consequently, at least one barrier or multiple hurdles resulting in a barrier need to be incorporated into the production process for products packaged using ROP. The incorporation of several sub-inhibitory barriers, none of which could individually inhibit microbial growth but which in combination provide a full barrier to growth, is necessary to ensure food safety.

Some products in ROP contain no preservatives and frequently do not possess any intrinsic inhibitory barriers (such as, pH, aw, or salt concentrations) that either alone or in combination will inhibit microbial growth. Thus, product safety is not provided by natural or formulated characteristics.

An anaerobic environment, usually created by ROP, provides the potential for growth of several important pathogens. Some of these are psychrotrophic and grow slowly at temperatures near the freezing point of foods. Additionally, the inhibition of the spoilage bacteria is significant because without these competing organisms, tell-tale signs signaling that the product is no longer fit for consumption will not occur. The use of one form of ROP, vacuum packaging, is not new. Many food products have a long and safe history of being vacuum packaged in ROP. However, the early use of vacuum packaging for smoked fish had disastrous results, causing a long-standing moratorium on certain uses of this technology.

Refrigerated Holding Requirements for Foods in ROP

Safe use of ROP technology demands that adequate refrigeration be maintained during the entire shelf-life of potentially hazardous foods to ensure product safety. Bacteria, with the exception of those that can form spores, are eliminated by pasteurisation. However, pathogens may survive in the final product if pasteurisation is inadequate, poor quality raw materials or poor handling practices are used, or post-processing contamination occurs. Even if foods that are in ROP receive adequate thermal processing, a particular concern is present at retail when employees open manufactured products and repackage them. This operation presents the potential for post-processing contamination by pathogens. If products in ROP are subjected to mild temperature abuse, *i.e.*, 5°-12°C (41°-53°F), at any stage during storage or distribution, foodborne pathogens, including Bacillus cereus, Salmonella spp., Staphylococcus aureus, and Vibrio parahaemolyticus can grow slowly. Marginal refrigeration that does not facilitate growth may still allow Salmonella spp., Campylobacter spp., and Brucella spp. to survive for long periods of time.

Recent published surveys indicate that refrigeration practices at retail need improvement. Some refrigerated products offered in convenience stores were found at or above 7.2°C (45°F) 50 per cent of the time; in several cases temperatures as high as 10°C (50°F) were observed. Delicatessen display cases have been shown to demonstrate poor temperature control. Foods have been observed above 10°C (50°F) and above 12.8°C (55°F) in several instances. Supermarket fresh meat cases appear to have a relatively good record of temperature control. However, even these foods can occasionally be found above 10°C (50°F).

Temperature abuse is common throughout distribution and retail markets. Strict adherence to temperature control and shelf-life must be observed and documented by the establishment using ROP. Information on temperature control should also be provided to the consumer. Currently these controls are not extensively used. Additionally, some commercial equipment is incapable of maintaining foods below 7.2°C (45°F) because of refrigeration capacity, insufficient refrigerating medium, or poor maintenance. Most warehouses and transport vehicles in U.S., distribution chains maintain temperatures in the 0°-3.3°C (32°-38°F) range. It must be assumed, however, for purposes of assessing risk, that occasionally temperatures of 10°C (50°F) or higher may occur for extended periods.

At retail, further temperature abuse must also be assumed. For instance, retail display cases can be as high as 13.3°C (56°F) for short periods and some refrigerated foods are provided no refrigeration for short periods of time. These realities point to the need for establishments to implement controls, such as buyer specifications, over refrigerated distribution systems so that better temperature control can be ensured.

Control of Clostridium botulinum and Listeria monocytogenes in Reduced Oxygen Packaged Foods

Recently, there has been an increased interest in vacuum packaging or MAP at retail using conventional refrigeration for holding. Refrigerated foods packaged at retail may be chilled either after they are physically prepared and repackaged, or packaged after a cooking step. In either case but primarily the latter, germination of Clostridium botulinum spores must be inhibited because spores are not destroyed by a heating step. Sanitary safeguards must be employed to prevent reintroduction of pathogens. Chief among these is Listeria monocytogenes.

Clostridum botulinum is the causative agent of botulism, a severe food poisoning characterised by double vision, paralysis, and occasionally death. The organism is an anaerobic spore-forming bacteria that produces a potent neurotoxin. The spores are ubiquitous in nature, relatively heat-resistant, and can survive most minimal heat treatments that destroy vegetative cells. Certain strains of C. botulinum (type E and non-proteolytic types B and F), which have been primarily associated with fish, are psychrotrophic and can grow and produce toxin at temperatures as low as 3.3°C (38°F). Other strains of C. botulinum (type A and proteolytic types B and F) can grow and produce toxin at temperatures slightly above 10°C (50°F). If present, C. botulinum could potentially grow and render toxigenic a food packaged and held in ROP because most other competing organisms are inhibited by ROP. Therefore, the food could be toxic yet appear organoleptically acceptable. This is particularly true of psychrotrophic strains of C. botulinum that do not produce tell-tale proteolytic enzymes. Because botulism is potentially deadly, foods held in anaerobic conditions merit regulatory concern and vigilance.

The potential for botulism toxin to develop also exists when ROP is used after heat treatments such as pasteurisation, or sous vide, processing of foods which will not destroy the spores of C. botulinum. Mild heat treatments in combination with ROP may actually select for C. botulinum by killing off its competitors. If the applied heat treatment does not produce commercial sterility, the food requires refrigeration to prevent spoilage and ensure product safety. For this reason, sous vide products are frequently flash frozen in liquid nitrogen and held in frozen storage until use.

There is a further microbial concern with ROP at retail. Processed products such as meats and cheeses which have undergone an adequate cooking step to kill L. monocytogenes can be contaminated when opened, sliced, and repackaged at retail. Thus, a simple packaging or repackaging operation can present an opportunity for recontamination with pathogens if strict sanitary safeguards are not in place.

Processors of products using ROP should be cautious if they plan to rely on refrigeration as the sole barrier that ensures product safety. This approach requires very rigourous temperature controls and monitored refrigeration equipment. If extended shelf-life is sought, a temperature of 3.3°C (38°F) or lower must be maintained at all times to prevent outgrowth of C. botulinum and the subsequent production of toxin.Listeria monocytogenes can grow at even lower temperatures; consequently, appropriate use-by dates must be established and readily apparent to the consumer. Since refrigeration alone does not guarantee safety from pathogenic microorganisms, additional growth barriers must be provided. Growth barriers are provided by hurdles such as low pH, aw, or short shelf life, and constant monitoring of the temperature. Any one hurdle, or a combination of several, may be used with refrigeration to control pathogenic outgrowth.

Design of Heat Processes for Foods in Reduced Oxygen Packages

Heat processes for sous vide or cook-chill operations should be designed so that, at a minimum, all vegetative pathogens are destroyed by a pasteurisation process. Special labeling of these products is necessary to ensure adequate warning to consumers that these foods must be refrigerated at 5°C (41°F) and consumed by the date required by the Code for that particular product.

The National Advisory Committee on Microbiological Criteria for Foods (NACMCF) chartered by the U.S. Department of Agriculture (USDA) and the Department of Health and Human Services (HHS) recently commented on the microbial safety of refrigerated foods containing cooked, uncured meat or poultry products that are packaged for extended refrigerated shelf-life and are ready-to-eat or prepared with little or no additional heat treatment. The Committee recommended guidelines for evaluating the ability of thermal processes to inactivate L. monocytogenes in extended shelf-life refrigerated foods. Specifically, it recommended a proposed requirement for demonstrating that an ROP process provides a heat treatment sufficient to achieve a 4 decimal log reduction (4D) of L. monocytogenes.

Other scientific reports recommend more extensive thermal processing. Thermal processes for sous vide practiced in Europe are designed to achieve a 12-13 log reduction (12-13D) of the target organismStreptococcus faecalis. It is reasoned that thermal inactivation of this organism would ensure destruction of all other vegetative pathogens.

Food manufacturers with adequate in-house research and development programmes may have the ability to design their own thermal processes. However, small retailers and supermarkets may not be able to perform the microbiological challenge studies necessary to provide the same level of food safety. If a retail establishment wishes to use an ROP process, microbiological studies should be performed by, or in conjunction with, an appropriate process authority or person knowledgeable in food microbiology who is acceptable to the regulatory authority.

Finally, if foods are held long enough, even under proper refrigeration, extended shelf-life may be a problem. A recent study on fresh vegetables inoculated with L. monocytogenes was conducted to determine the effect of CAP on shelf life. The study found that CAP lengthened the time that all vegetables were considered acceptable, but that populations of L. monocytogenes increased during that extended storage.

Consumer Handling Practices and In-Home Refrigeration Temperatures

Extended shelf-life provided by ROP is cause for concern because of the potential for abuse by the consumer. Consumers often can not, or do not, maintain adequate refrigeration of potentially hazardous foods at home. Foods in ROP that are taken home may not be eaten until enough time/temperature abuse has occurred to allow any pathogens present to increase to levels which can increase the chance of illness.

Under the best of circumstances home refrigerators can be expected to range between 5° and 10°C (41°-50°F). One study reported that home refrigerator temperatures in 21 per cent of the households surveyed were 10°C (50°F). Another study reported more than 1 of 4 home refrigerators are above 7.2°C (45°F) and almost 1 of 10 are above 10°C (50°F). Thus, refrigeration alone cannot be relied on for ensuring microbiological safety after foods in ROP leave the establishment.

Consumers have come to expect that certain packages of foods would be safe without refrigeration. Low-acid canned foods have been thermally processed, which renders the food shelf-stable. Retort heating ensures the destruction of C. botulinum spores as well as all other foodborne pathogens. Yet consumers may not understand that most products that are packaged in ROP are not commercially sterile or shelf-stable and must be refrigerated.

A clear label statement to keep the product refrigerated must be provided to consumers. The use of ROP has been extensively studied by regulators and the food industry over the past several years. Recommendations have been adapted from the Association of Food and Drug Officials "Retail Guidelines - Refrigerated Foods in Reduced Oxygen Packages" and New York State Department of Agriculture and Markets "Proposed Reduced Oxygen Packaging Regulations.

" As provided in the Food Code, some ROP operations may be conducted under provision 3-502.12 Reduced Oxygen Packaging, Criteria. Food that is packaged by an ROP method under these provisions is considered safe while it is under the control of the establishment and, if the labeling instructions are followed, while under the control of the consumer.

Safety Barrier Verification

The safety barriers for all processed foods held in ROP at retail must be verified in writing. This can be accomplished through written certification from the product manufacturer. Independent laboratory analysis using methodology approved by the regulatory authority can also be used to verify incoming product and should be used to verify the barriers in a product that is packaged within the establishment by an ROP method.

It should be noted that the Association of Food and Drug Officials (AFDO) guidelines recommend that laboratory analysis be conducted by official methods of the Association of Official Analytical Chemists (AOAC). The multiple barrier or hurdle efficacy should be validated by inoculated pack or challenge studies. A product should be tested under abuse temperatures to demonstrate product safety during the food's shelf life.

Any changes in product formulation or processing procedures are cause for notification of the regulatory authority and a required approval of the revised ROP process.

A record of all safety barrier verifications should be updated every 12 months. This record must be available to the regulatory authority for review at the time of inspection.

USDA Process Exemption

Meat and poultry products cured at a food processing plant regulated by the U.S. Department of Agriculture using substances specified in poultry products are exempt from the safety barrier verification requirements.

Recommendations for ROP Without Multiple Barriers

- *Employee Training:* If ROP is used, employees assigned to packaging of the foods must have documented proof that demonstrates familiarity with ROP guidelines in this Annex and the potential hazards associated with these foods. At the discretion of the regulatory authority, a description of the training and course content provided to the employees must either be available for review or have prior approval by the regulatory authority.
- *Refrigeration Requirements:* Foods in ROP that have only one barrier, *i.e.*, refrigeration, to C. botulinum must be refrigerated to 5°C (41°F) or below and marked with a use-by date within either the manufacturer's labeled use-by date or 14 days after preparation at retail, whichever comes first. Alternatively, foods packaged by ROP may be kept frozen if freezing is used as the declared primary safety barrier. Any extension of shelf life past 14 days will require a further variance that considers lower refrigeration temperatures. Foods that are intended for refrigerated storage beyond 14 days must be maintained at or below 3°C (38°F).
- *Labeling: Refrigeration Statements*: All foods in ROP which rely on refrigeration as a barrier to microbial growth must bear the statement "Important - Must be kept refrigerated at 5°C (41°F)" or "Important - Must be kept frozen," in the case of foods which rely on freezing as a primary safety barrier. The statement must appear on the principal display panel in bold type on a contrasting background. Foods held under ROP which have lower refrigeration requirements as a condition of safe shelf life must be monitored for temperature history and must not be offered for retail sale if the temperature and time specified in the variance are exceeded.
- *Labeling: "Use-by date"*: Each container of food in ROP must bear a "use-by" date. This date cannot exceed 14 days from retail packaging or repackaging without a further variance granted by the regulatory authority. The date assigned by a repacker cannot extend beyond the manufacturer's recommended "pull date" for the food. The "use-by" date must be listed on the principal display panel in bold type on a contrasting background. Any label must contain a combination of a "sell-by" date and use-by instructions which makes it clear that the product must be consumed within 14 days of retail packaging or

repackaging, as an acceptable alternative to a 14 day "use-by" date, *i.e.*, for product packaged on November 1, 1999 - "Sell by November 10, 1999" - use within 4 days of sell-by date. Foods that are frozen before or immediately after packaging and remain frozen until use should bear a "Keep frozen, use within 4 days after thawing" statement.

Foods Which Require a Variance

- Processed fish and smoked fish may not be packed by ROP unless establishments are approved for the activity and inspected by the regulatory authority. Establishments packaging such fish products, and smoking and packing establishments, must be licensed in accordance with applicable law. Caviar may be packed on the premises by ROP if the establishment is approved by the regulatory authority and has an approved scheduled process established by a processing authority acceptable to the regulatory authority.
- Soft cheeses such as ricotta, cottage cheese, cheese spreads, and combinations of cheese and other ingredients such as vegetables, meat, or fish at retail must be approved for ROP and inspected by the regulatory authority.
- Meat or poultry products which are smoked or cured at retail, except that raw food of animal origin which is cured in a USDA-regulated processing plant, or establishment approved by the regulatory authority to cure these foods may be smoked in accordance with approved time/temperature requirements and packaged in ROP at retail if approved by the regulatory authority.

Hazard Analysis and Critical Control Point (HACCP) Operation

All food establishments packaging food in a reduced oxygen atmosphere must develop a HACCP plan and maintain the plan at the processing site for review by the regulatory authority.

For ROP operations the plan must include:

- A complete description of the processing, packaging, and storage procedures designated as critical control points, with attendant critical limits, corrective action plans, monitoring and verification schemes, and records required;
- A list of equipment and food-contact packaging supplies used, including compliance standards required by the regulatory authority, *i.e.*, USDA or a recognised third party equipment by the evaluation organisation such as NSF International;
- A description of the lot identification system acceptable to the regulatory authority;
- A description of the employee training programme acceptable to the regulatory authority;

- A listing and proportion of food-grade gasses used; and
- A standard operating procedure for method and frequency of cleaning and sanitising food-contact surfaces in the designated processing area.

Precautions Against Contamination at Retail

Only unopened packages of food products obtained from sources that comply with the applicable laws relating to food safety can be used to package at retail in a reduced oxygen atmosphere. If it is necessary to stop packaging for a period in excess of one-half hour, the remainder of that product must be diverted for another use in the retail establishment.

Disposition of Expired Product at Retail

Processed reduced oxygen foods that exceed the "use-by" date or manufacturer's "pull date" cannot be sold in any form and must be disposed of in a proper manner.

Dedicated Area/Restricted Access

All aspects of reduced oxygen packaging shall be conducted in an area specifically designated for this purpose. There shall be an effective separation to prevent cross contamination between raw and cooked foods. Access to processing equipment shall be restricted to responsible trained personnel who are familiar with the potential hazards inherent in food packaged by an ROP method. Some ROP procedures such as sous vide may require a "sanitary zone" or dedicated room with restricted access to prevent contamination.

FREEZING AND COLD STORAGE OF FISH

It is perfectly feasible to keep fresh fish for many months without perceptible change in the eating quality by rapidly freezing it soon after catching and then storing it at a suitably low and constant temperature. With this method of preservation, the thawed product is virtually indistinguishable from the best fresh fish. Freezing has revolutionised the fish processing industry in Britain since the Second World War.

FREEZING FISH

Heat is removed from the fish in the freezing process either by surrounding the fish with a stream of cold air, by placing the fish in contact with a cold surface or by spraying with certain liquid refrigerants. Three main types of freezing plant are used that employ these techniques, the air blast freezer, the plate freezer and the immersion freezer. The air blast freezer is essentially a tunnel in which a fast-moving stream of very cold air is blown over the fish, which are placed on trolleys or on a moving belt. The air is usually at a temperature of -30 to -40°C and moving at about 5 m/s. The air blast freezer is most suitable for a wide range of sizes of fish, and for products of irregular shape.

The plate freezer is more compact than the air blast freezer, and is most useful for handling fish products that are uniform in thickness and that have a reasonably flat surface which can make good contact with the cold plates. Two versions of the plate freezer are in general commercial use, the horizontal and the vertical types. The horizontal plate freezer, used mainly in land installations, handles many of the catering and retail fish products that are already packed in cartons prior to freezing. Trays of packs are slid between pairs of horizontal plates, the plates are closed tightly on to the packs by hydraulic pressure to make good contact, and a cold refrigerant is circulated through serpentine passages within the plates; a retail pack 3 cm thick takes about an hour to freeze.

The vertical plate freezer, which was originally designed in the 1950s for use on fishing vessels, is employed for freezing large blocks of whole fish. The fish are packed between pairs of plates, usually without any wrappings, and the plates moved slightly towards each other to compact the block and ensure good contact. Liquid refrigerant is circulated through the serpentine passages within the plates until the fish are frozen; the complete process for a block of whole cod 10 cm thick takes about 4 hours including loading and unloading time, with refrigerant at -40°C.

The immersion freezer is not used as much as the plate and blast freezers in the British fish industry; the main difficulty is the limited choice of liquids that are both good refrigerants and suitable for use in direct contact with foods. Brines are sometimes used, but the fish may take up too much salt. Liquid nitrogen is used successfully in one type of immersion freezer, which subjects the product to a spray of liquid nitrogen as it passes through a tunnel.

The freezing process cannot improve the quality of fish; therefore the best frozen products are those made from first class raw material. This applies particularly to whole fish which, after thawing, are likely to be subjected to further processing. Whole iced cod, for example, when frozen not later than 3 days after catching can on thawing be treated in the same way as very fresh fish, but cod that has been delayed longer than this, or has been kept uniced, is unlikely after freezing and thawing to yield fillets of a high quality. Whilst some species, flatfish for example, can be kept in ice a little longer than cod before freezing without impairing the quality of the thawed product, others like haddock and hake do not keep so well. Fillets of most white fish species can be taken from whole fish 5-6 days in ice and frozen to give a high quality thawed product.

The freezing process should always be completed as rapidly as possible, not only to increase output of the equipment, but also to reduce the time in which bacteria and digestive juices are still able to attack the fish; bacterial action ceases below about -10°C, and the activity of enzymes is reduced as the temperature falls. There are marked changes in texture and flavour when fish is frozen very slowly at temperatures only a little below 0°C. The final temperature of fish being frozen should be that at which it is to be stored, namely

-30°C; this ensures that the frozen product imposes no extra heat load on the cold store since this is designed only to keep the product cold and not to freeze it.

Freezing at Sea

On freezer trawlers handling whole fish, the catch is brought in over the stern, discharged to an enclosed processing deck, gutted by hand or machine, washed and bled in cold sea water and conveyed forward to the freezers, where blocks of about 50 kg are produced in vertical plate freezers and transferred to cold storage rooms running at -30°C. These ships make voyages of 4-8 weeks, carry a crew only two or three more in number than their counterparts handling iced fish, and bring back about 500 tonnes of frozen cargo for transfer to shore cold stores at the ports. Factory trawlers that fillet the catch before freezing are less numerous than trawlers which freeze whole fish.

On the former the gutted fish are filleted mainly by machine, then the fillets are packed into blocks of about 10-15 kg and frozen in horizontal plate freezers before being transferred to the ship's cold store. Compared to the situation with a trawler freezing whole fish, the crew has to be larger to cope with factory operation at sea, and the quality of fillets frozen at sea as opposed to whole fish frozen at sea depends much more on careful handling and processing, but on the other hand the storage capacity of the ship in terms of frozen edible portion is considerably increased. It is likely that the distant water fishing fleet will gradually change over completely from the stowage of wet fish in ice to the freezing of fish at sea.

Cold Storage of Fish

Temperature of storage is the most important single factor affecting the storage life of frozen fish. Almost all cold stores for fish that have been built in recent years are designed to operate at -30°C, at which temperature the products will keep in first class condition for several months. Lean fish, such as cod and haddock, when stored for long periods at too high a temperature or in fluctuating temperatures can have marked changes in texture and flavour when thawed out. The thawed flesh may feel rubbery and appear dense white instead of translucent. After cooking they are found to be tough and fibrous or stringy.

Other causes of change in cold stored fish are dehydration and oxidation. Dehydration is kept to a minimum either by glasing unwrapped frozen fish before storing them, that is covering the surface with a skin of ice by dipping them quickly in cold water, or by packaging the fish in a material that is a good barrier against the passage of water vapour, for example polyethylene film. Fatty fish like herring are particularly prone to absorb oxygen from the air and so become rancid; these are therefore wrapped in a material that forms a good oxygen barrier. In addition the space between the package and the contents may be evacuated to reduce even further the risk of rancidity. The ideal packaging material for fish products is often a laminated film combining the desired properties of two or more plastics. The cold storage chain is maintained in distribution by the use of insulated, refrigerated vehicles or containers, both of which operate at about -20°C, and frozen food cabinets, again at -20°C, in shops and catering premises.

Thawing Frozen Fish

The growth of quick freezing in the fish industry, and particularly the production of large blocks of whole fish for subsequent processing, has made necessary the development of thawing plant. There are two main types of equipment, those in which the fish are heated in a warm air stream or in warm water, and those which directly use heat generated by electricity. The method most used for large blocks of sea frozen fish is air blast thawing; the fish are conveyed through a stream of moist moving air at about 20°C until they are thawed enough to permit filleting or other processing. The fish temperature should never exceed 20°C.

DRIED AND SALTED FISH

Bacteria and moulds generally cannot grow in the absence of water and hence drying can be used as a means of preservation. Salt, if present in sufficient strength, will slow down or prevent bacterial spoilage of fish. Drying or salting, or a combination of both, have been used in the fish industry for centuries but nowadays only a very small proportion of the catch in Britain is processed by these methods.A few companies still make dried salted fish from cod and related species, by heading and splitting the fish, removing most of the backbone, and stacking the fish in piles with layers of salt between them. The juices withdrawn from the fish by the salt are allowed to run away and, after frequent restocking over a period of months, the water content is further reduced by drying the fish in a heated chamber until the moisture content is somewhere between 10 and 30 per cent.

Some herring are pickle cured, particularly in north-east Scotland and the Shetlands, although this export trade is a mere shadow of what it was earlier this century. The whole herring, having been lightly sprinkled with salt while awaiting processing, are first gibbed by hand or machine, that is the gills, long gut and stomach are removed. They are then packed in barrels with a layer of salt on each layer of herring until the barrels are full. After a day or two, when the herring have shrunk appreciably, the barrels are topped up with further layers of herring and salt, the lids are fitted and the barrels then laid on their sides for 8-10 days. After this period the barrels are up-ended, the lids removed, and the blood pickle from the upper half drained off through the bung-hole. To make ready for storage the barrels are finally topped up with fish, the lids replaced and blood pickle poured through the bung-holes until all spaces are filled. Klondyking is the name given to another method of preserving ungutted herring using salt. The name is thought to originate from a method developed about the same time as the famous Gold Rush of 1897. The herring, contained in baskets, are sprinkled with coarse salt, approximately 175 kg of fish treated with about 10 kg of salt, as they are tipped into a wooden box. Ice is then put on this mixture offish and salt, about 100-125 kg of ice to every 175 kg of mixture. Herring treated in this way could be kept in edible condition for approximately 1 week.

CANNED FISH

The canned fish industry in Britain is a small one, and the range of products is confined mainly to herring, sprats and pilchards packed in either tomato sauce or vegetable oil. The process for herring in tomato sauce is typical. First the fish are nobbed by machine, that is the head and gut are removed. They are then immersed in saturated brine for up to 30 minutes and packed by hand into oval cans holding 200 g of fish. The tomato sauce is added, the can lids are lightly clipped on and the cans exhausted in steam for 10-15 minutes. The object of exhausting is to produce a partial vacuum in the headspace of the can which is not filled either with solids or liquids. The cans are then sealed, washed, and heat processed in steam at 115°C for 55 minutes. After cooling, the cans are stored for about a month, labelled and packed in outer cartons for dispatch. The heat processing stage is critical, and is designed to inactivate all bacteria and enzymes present and in particular to destroy any harmful organisms. There are a number of imported canned fish products that could be manufactured in this country from British-caught fish, and the canning industry is examining the possibilities of expanding their range.

Bibliography

A L Bhatia: *Biochemistry and Endocrinology*, Indus Valley, Delhi, 2002.

A.D. Dholakia: *Delicious Seafood Recipes*, Daya Publication, Delhi, 2010.

A.K. Bakshi, V.K. Joshi, Devina Vaidya and Savita Sharma: *Food Processing and Preservation (2 Parts-Set)*, Jagmander Book Agency, Delhi, 2013.

Akhilesh Sharma: *Biochemistry*, RBSA Publication, Delhi, 2006.

Arnold C. Long: *Advances in Seafood Biochemistry: Composition and Quality*, Cyber Tech Publication, Delhi, 2009.

B.B. Buchanan, Wilhelm Gruissem and Russell L. Jones: *Biochemistry and Molecular Biology of Plants*, I.K. International, Delhi, 2007.

Bimla Singh: *Biochemistry and Genomic Revolution*, Vista International Publication, Delhi, 2006.

Chanderlekha Goswami: *Biochemistry and Instrumentation*, Manglam Publication, Delhi, 2011.

H H Huss; L Ababouch and L Gram: *Assessment and Management of Seafood Safety and Quality*, Daya Publication, Delhi, 2007.

J C Blackstock: *Biochemistry*, Viva Books, Delhi, 2008.

John K. Joseph: *Biochemistry*, Campus Books, Delhi, 2006.

L. Veerakumari: *Biochemistry*, MJP Publication, Delhi, 2004.

Mohd. Azharul Haque: *Biochemistry*, Jnanada Prakashan, Delhi, 2014.

Mridula D., R.T. Patil and M.R. Manikantan: *Food Processing Technologies Co-Product Utilization and Quality Assurance*, Satish Serial Publishing House, Delhi, 2012.

N. Sarath Chandra Bose: *Biochemistry : A Practical Manual*, Pharma Med Press, Delhi, 2010.

N.L. Choudhary and Anjana Singh: *Food Processing Biotechnology Applications*, Oxford Book Company, Delhi, 2012.

Neelam Khetarpaul: *Food Processing and Preservation*, Daya Publication, Delhi, 2005.

Pradeep Sharma: *Biochemistry and Organisation of Cells*, RBSA Publication, Jaipur, 2006.

R.K. Verma: *Food Processing Management*, Pearl Books, Delhi, 2011.

Ram Naresh Mahaling: *Basics of Biochemistry*, Anmol Publication, Delhi, 2008.

S Banerjee: *Biochemistry and Biotechnology*, Dominant Publication, Delhi, 2009.

S.S. Purohit: *Biochemistry : Fundamentals and Applications*, Agrobios Publication, Delhi, 2009.

Subha Ganguly: *Food Processing and Quality Control*, Narendra Publishing House, Delhi, 2014.

V.K. Yadav and Neelam Yadav: *Biochemistry and Biotechnology : A Laboratory Manual*, Pointer Publication, Delhi, 2007.

Veena: *Biochemistry*, Sonali Publication, Delhi, 2008.

William H. Elliott and Daphne C. Elliott: *Biochemistry and Molecular Biology*, Oxford University Press, Delhi, 2005.

Y.S. Reddy: *Food Processing and Bioactive Compounds*, Gene Tech Books, Delhi, 2006.

Index

H

I

L

M

N

O

P

R

S

T

U

V

W